CAMBRIDGE COUNTY GEOGRAPHIES

General Editor: F. H. H. GUILLEMARD, M.A., M.D.

T0174431

HUNTINGDONSHIRE

Cambridge County Geographies

HUNTINGDONSHIRE

by

THE REV. W. M. NOBLE

With Maps, Diagrams and Illustrations

Cambridge :

at the University Press

1911

CAMBRIDGE UNIVERSITY PRESS
Cambridge, New York, Melbourne, Madrid, Cape Town,
Singapore, São Paulo, Delhi, Mexico City

Cambridge University Press
The Edinburgh Building, Cambridge CB2 8RU, UK

Published in the United States of America by Cambridge University Press, New York

www.cambridge.org
Information on this title: www.cambridge.org/9781107645967

© Cambridge University Press 1911

This publication is in copyright. Subject to statutory exception
and to the provisions of relevant collective licensing agreements,
no reproduction of any part may take place without the written
permission of Cambridge University Press.

First published 1911
First paperback edition 2013

A catalogue record for this publication is available from the British Library

ISBN 978-1-107-64596-7 Paperback

This publication reproduces the text of the original edition of the
Cambridge County Geographies. The content of this publication has
not been updated. Cambridge University Press has no responsibility
for the accuracy of the geographical guidance or other information
contained in this publication, and does not guarantee that such
content is, or will remain, accurate.

CONTENTS

ILLUSTRATIONS

MAPS

The illustrations on pp. 7, 19, 50, 51, 53, 81, 82, 92, 93, and 125, are from photographs by Mr Freeman of Ramsey: those on pp. 70, 124, and 127, from photographs by Mr Hendry of Godmanchester: that on p. 132 from a photograph by Mr J. H. Heathcote: that on p. 121 by Messrs Maddison and Hinde of Huntingdon: those on pp. 21, 22 by Mr Percy Slater of Sawtry: that on p. 61 by Mr E. H. H. Hancox of Nacton: those on pp. 11, 16, 112 by Mr A. Wright: that on p. 36 by Mrs E. A. Brawn: that on p. 75 by Dr Guillemard: that on p. 139 by Mr W. H. Hayles: that on p. 130 is given by kind permission of the Master and Fellows of Magdalene College, Cambridge: that on p. 127 by kind permission of the Earl of Sandwich: that on p. 124 by kind permission of Archdeacon Vesey: those on pp. 84, 102 by kind permission of the late Earl of Carysfort: those on pp. 3, 14, 24, 25, 34, 39, 55, 56, 57, 58, 96, 97, 101, 104, 106, 107, 108, 113, 116, 117, 134, 140, 143, 146, are by Messrs Frith and Co.

1. County and Shire. The Word Huntingdonshire: Its Origin.

From very early times, as soon as a land had inhabitants, names were of necessity given to places and to persons, in order to distinguish one place from another and one person from another.

In many cases the original names have passed away and others have taken their place. A successful invasion, for example, might sweep away inhabitants, names, and even language, and the new comers would proceed to give fresh names to the places they had captured. Our country was once called Albion, the White Island, from its chalk cliffs facing the continent of Europe, but that name is scarcely heard now, and even Britain, another name given to it, is not now used for this particular part of the British islands, but only as a general name embracing more than this island itself. We know it as England, that is Angleland—the land of the Angles or Engles—and we shall see later how this name came to be applied to the country.

Names of places are very often closely connected with those of persons, sometimes a person is named after a place, sometimes a place after a person. In early days an individual had but one name; a man named Henry, for example, might leave his native village of Ellington, and would become Henry of Ellington, and hence Henry Ellington. Or sometimes the man's trade gave him his name—surname as we call it—and John the smith, and Philip the tailor, became known as John Smith and Philip Taylor. Or perhaps the name came in a different way. William, whose father was named John, was spoken of as William, the son of John, soon abbreviated to William Johnson. Names of places, too, were often very appropriate and conveyed a meaning, thus "ey" was an old word of Scandinavian origin, meaning "island," so Ramsey conveys to us the fact that Ramsey is, or once was, an island. Similarly the ending of a name in "ford" suggests that the place was near a fordable river—thus Hemingford is the ford near which the Hemmings or sons of Hemma lived. "Den' or "dene" means a hollow, so we find Denton in a hollow.

Names of places, then, are often very closely connected with those of persons, and this is the case as regards the name of Huntingdon. There can be little doubt that one Hunta, a hunter of such note that he had acquired his name from his skill in that calling, had possessions and his residence at or near the site of the present town of Huntingdon. In time, the hill on which stood Hunta's residence came to be called Huntandun, or Hunta's Hill; the name gradually taking a wider signification and being applied to the whole county which we now know as Huntingdonshire. Who Hunta was or whence he came is entirely unknown, but as he was a man of sufficient note to give his name not only to his residence but to the

Market Hill, Huntingdon

1—2

district round it, he may have ruled or owned that part of England which is now Huntingdonshire.

To-day we have England divided into counties, and it will be noticed that the names of some of these end in "shire"—as Cambridgeshire and Huntingdonshire—and some, as Kent and Essex, do not. We naturally ask the reason for this difference. Let us consider the meanings of the two words "shire" and "county." England was not in early days one country under a single king, but consisted of many small kingdoms, and even the boundaries of some of these were often changed as one kingdom or another obtained ascendency. The kingdoms themselves were frequently subdivided, the subdivisions being ruled by ealdormen or earls under the king; these subdivisions were called "shares" or "shires," that is, parts of a kingdom, so we have Huntingdonshire, the "shire" or "share" under Hunta. The counties of Kent and Essex, however, are the original undivided kingdoms; they are not "shires" or "shares" "shorn off" (for the word is the same) from any larger dominions.

After the Norman Conquest, Counts were appointed to govern parts of England. The district governed by a Count was called his county; this frequently corresponded exactly with the area of a dispossessed Saxon Earl, so the district previously called a shire now came to be called a county, and this latter name applied also to other areas, as to Kent. In time such divisions came to be called county or shire indiscriminately.

Alfred the Great is usually credited with the division of England into counties, but that has been a very gradual work. Even at the present day changes are being made in the boundaries of counties; that of Huntingdonshire has been changed, and a new County of London has been formed.

It is worth noting how compact the kingdom of East Anglia was—almost a square—and how ready the early inhabitants were to take advantage of natural boundaries. The northern and eastern limits of East Anglia were formed by the sea, and the invaders, having occupied the seaboard, pushed up the Ouse on the west and the Stour on the south, advancing as far inland as possible and gradually extending till all South Hunts was occupied, availing themselves, in short, of much the same natural boundaries as did their predecessors, the Iceni.

It is interesting to note, too, that many of our parishes are bounded by streams, large or small, for parishes were formed out of manors, portions of property held by some individual owner, who had as the boundary of his estate some immovable and natural landmark. The Saxon earl's estate or manor was a kingdom on a small scale.

A glance at a map of the United States will show that natural boundaries had little weight with those who formed the plan of the later States. It is true that these are sometimes bounded by rivers or mountains, but far oftener by straight lines arbitrarily drawn, in accordance with smaller subdivisions, which are always, where possible, rectangular. We can see, in short, that these countries have not grown slowly and naturally, as has been the case in England, but have been arbitrarily and simultaneously parcelled out by the hand of man.

2. General Characteristics of the County. Its Position.

Huntingdonshire is an inland county in the south-eastern midlands of England, lying south-west of the Wash, from which its nearest point is distant about 30 miles. King's Lynn is the nearest seaport. As Huntingdonshire produces neither coal nor iron, its manufactures are naturally on a small scale, and the energies of its inhabitants are devoted principally to agriculture. Huntingdonshire was anciently forest, that is uncleared land, and so specially suited for hunting; here and there would be clearings with small cultivated spots —the homes of a sparse population. Its deforestation took place in the reign of Edward I, and doubtless had the effect of increasing the productiveness of the county; indeed in the early part of the nineteenth century much corn was exported, and, as the fen drainage became more and more effective, additional land was brought into cultivation, till the agricultural produce of the county has become much above the average in proportion to its acreage.

The southern and western portions of Huntingdonshire present just such a picture as one may see in many midland counties of England—pleasantly diversified hill and dale, with no hills of remarkable height and no very deep valleys, but each valley with its stream making its way to the Ouse or the Nene. This part of the county is locally known as the "highland," in contradistinction to the fen, which will be mentioned later. Part of the highland is arable and produces the usual crops, part is devoted to dairy farming, but as we reach the northern part of the county we find a soil not so good as in the southern part, and the tall chimneys of the brickyards tell us of the industry that is carried on in this part of the county.

The Mere Mill, River Nene

There is a great change when we enter the north-eastern portion of the county. As we approach it from the hills on a bright summer day, there appears what looks like an absolutely level plain, dotted with a few homesteads surrounded by clumps of trees, and here and there a cottage, while the straight dykes, cut beside the fields, in which the water glistens in the sunlight, make it look like some floor ruled with silver lines. Once we are in it, we can fancy ourselves in Holland; the roads are usually perfectly straight for miles, with a ditch—or dyke as it is termed—on either side and every field is bounded in the same way by a perfectly straight dyke on each side. These are all part of the system of drainage, all those on each level being connected, and each level with a higher or lower one. Hedges there are practically none; dykes take their place and serve the double purpose of draining the land and dividing the fields.

In the fen is very little grass land, the soil being too light for pasture, but abundant crops of mangolds, celery, potatoes, and other roots are grown, and these often reach a large size—while heavy wheat crops with long straw alternate with the above.

3. Size. Shape. Boundaries. Detached Portions.

Huntingdonshire is one of the smallest counties in England, Rutland and the old county of Middlesex alone being smaller. It is about 30 miles in length from north to south and 23 in breadth from east to west, and is approximately diamond-shaped, but in the northern part is a considerable projection westward which makes the west side very irregular. Northamptonshire forms the northern and north-western boundary, Bedfordshire the south-western, and Cambridgeshire the north-eastern, eastern, and south-eastern.

Few counties have more irregular boundaries than Huntingdonshire, and few are so little dependent upon natural features to form them. Starting from Peterborough which stands at the junction of the three shires of Northampton, Cambridge, and Huntingdon, and working westward, we find that the river Nene does indeed form the limit of the county for some 12 miles or so, but it ceases to do so at Elton. Here the boundary turns east, formerly passing actually through the house at Elton Hall, though lately it has been slightly shifted. From here its course, though roughly southerly, is extremely irregular. At Covington is the Three Shire Stone, marking the limit of Bedfordshire to the north. From this point the general direction of the boundary is south-eastward. At one time the boundary was complicated still further here by a detached portion—Swineshead, which will be referred to later—lying isolated in Bedfordshire. For about half a mile the Kym, or Kim, forms the limit just below Kimbolton and again for about the same distance in the neighbourhood of St Neots, immediately before its junction with the Great Ouse. Near Waresley the southern limit of the county is reached, and the third side of the diamond is entered upon, the general trend being north-east. At first, however, the line separating the county from Cambridgeshire is extraordinarily sinuous, two portions of the latter projecting as peninsulas far into Huntingdonshire. Reaching Ermine Street the line leads south again along it for a short distance and thus brings in Hilton with its quaint old-world green and curious maze, and crossing the Via

Devana reaches the Ouse again at Holywell near St Ives. For nearly six miles to the end of the third side of the diamond the river is the boundary until Earith is reached. Here the Bedford River runs in, and here begins the last section running north-east through the great flats of

The Maze, Hilton

Fenland and hardly passing anything larger than a farm or small hamlet until Peterborough is once more reached.

The Local Government Act of 1888 has made some alterations in county boundaries. By this Act, Swineshead has, for administrative purposes, been transferred to Bedfordshire. Tilbrook has been transferred from Bedfordshire to Huntingdonshire, while parts of Luddington, Lutton, and Thurning—formerly in Huntingdonshire— have been transferred to Northamptonshire; that part of Winwick, which was in Northamptonshire, to Huntingdonshire; part of Papworth St Agnes from Huntingdonshire to Cambridgeshire; and part of the soke of Peterborough

The Nene near Elton

from Huntingdonshire to Northamptonshire; making the area of the administrative county slightly smaller than that of the geographical county.   The area of the county before the Local Government Act was 366 square miles, but, according to the latest returns, the geographical area is 234,218 acres, and that of the administrative county 233,984 acres.

As in the case of other counties, there are some parishes on the border which are partly in Huntingdonshire and partly in other counties. For example, Great Catworth has some small portions in Northamptonshire; Midloe, which is locally in the parish of Southoe, is in Graveley, Cambridgeshire; while of the 2404 acres of land and water which form the parish of Stanground, no less than 1117 acres are in Cambridgeshire. On the other hand, the parish of Swineshead is entirely detached from the rest of the county of Huntingdon and is surrounded by Bedfordshire. The reason of this is that the parish stood within the great manor of Kimbolton, once the property of Earl Harold—the last Saxon king of England—and, as part of this manor was held to be in Huntingdonshire though separated from the rest of the county by about half a mile at the nearest point, and, though King Harold's lands were granted to different persons by the Conqueror, Swineshead and Stonely both passed to Fitz-Piers and both remained in the county of Huntingdon. We find that at the time of the Domesday survey land in Kimbolton and in Swineshead was held by William de Warene, and his influence would naturally be used to keep his property in the same county.

4. Surface and General Features. The Fens and Meres.

The general surface of the county is low, the highest point being at Keyston toll-house, which is between Huntingdon and Thrapston, at almost the extreme western part of the county.   This is 245 feet above sea-level, while St Ives—six miles east of Huntingdon—is only 26 feet, and Huntingdon 45 feet above the sea.

The surface of the county in its southern and western parts consists of undulating ground.   Though Huntingdon-shire cannot claim the wild beauty of Cumberland and Westmorland, nor the charming variety of the scenery of Devon and Cornwall, yet a ride in early autumn through the districts just mentioned shows the traveller a type of undulating land characteristic of the scenery of the Mid-lands of England, with its hills and valleys, its trees and brooks, its well-kept hedgerows bounding fields rich with waving corn, and meadows with herds of cattle and flocks of sheep—a very pleasing picture of rural England.

The north-eastern portion of the county is entirely different both in appearance and in the nature of its soil, consisting of fen, which, as we have seen, is almost perfectly flat, with few hedges or trees, but intersected by many rivers and dykes of different breadth.   It does not require a great effort of the imagination to picture the fens as a sea or inland lake, the "highland" round it forming the edge and sloping down to it, which is exactly as the fens were, at times, centuries ago.   Here and there some parts of the surface are somewhat higher, having been islands in the midst of the surrounding water and fen.   These elevations are formed by inequalities in the Oxford Clay—the substratum of the whole county—which here rises almost to the surface, covered only by drift.   The town of Ramsey is an instance of this, standing on what was once an island, rising above the surrounding fen.

The Ouse at Huntingdon

Yet, in spite of all that may be said as to the monotony of a level surface stretching in some directions as far as the eye can see, the fens of to-day have a beauty all their own: the waving corn, in its rich luxuriance, growing to a height unsurpassed anywhere in England, while the variety of tint given by the intervening patches of potatoes, beans, yellow-flowered mustard, cole seed, and other crops make a pleasing picture suggestive of comfort and abundance.

The fen district of Huntingdonshire consists of 35,000 acres, or about one-eighth of the whole county.

A mere level plain presents perhaps little of interest to the onlooker, but when he finds a piece of unreclaimed fen-land, with its growth of coarse grasses, bulrushes, and small bushes, and learns that these grow upon a soil that is almost all formed of decayed vegetation; when he hears, too, that underlying this is peat, which can be dug for fuel; that below this layer of peat lies silt, sand, or clay; that below this again occurs yet another layer of peat; and that this in its turn rests on a lower bed of clay, it is very natural for him to conclude that great changes must have occurred, and to ask how these changes were brought about. The question which presents itself is—How did the fens and meres develope into what has been termed the "Golden Plain of England"?

The Fens and their Drainage.

The level tract of the Fenland, the largest plain in Britain, reaches from Lincoln on the north almost to Cambridge on the south, and from Stamford on the west

The Huntingdon Fenland

to Brandon on the east, and contains an area of 750,000 acres. It received the name of the Bedford Level from the Earl of Bedford, whose association with the reclamation of the fens will presently be referred to, and it is divided into three parts called the North Level, South Level, and Middle Level. Part only of this plain is covered by peat.

The area of fen land in the county of Huntingdon is given, as mentioned above, as 35,000 acres—some authorities, however, say 44,000. The whole of this is covered by peat and is in the Middle Level, with the exception of a small portion in the South Level.

Miller and Skertchley in *Fenland, Past and Present* thus describe the unreclaimed fen:—"A vast open plain, covered, for the most part, with deep sedge, dotted with thickets of alder and willow, abounding in shallow lakes, temporary and permanent, and overflowed in its lowest parts, nearly, if not every winter."

While there have been great changes in the Fenland in the last 2000 years, it would appear that in the middle ages the fenmen during the winter months could often travel from place to place without following the course of the river, for all the fen was one vast lake. As the spring advanced and the waters sank, coarse grasses grew, and most of the fen became passable for foot-passengers, with pools here and there and some lakes of considerable size, for the land was not a dead level, but in places were depressions of several feet, forming permanent lakes. We may gather from the above that what was often a lake in winter became in summer a rough plain covered with sedge, reeds, alders, and willows as we find it in some parts to-day.

Let us now proceed to consider how the silt and peat came to be in their present positions, or in other words how the fens were formed. A glance at the map will show that the Fenland receives the drainage of several counties, the Huntingdonshire district collecting the waters from most part of the "highland" portion of the county. We notice, too, that owing to the flatness of the country the rivers are very sluggish. Now something of this sort must have happened : at certain times the rivers over-flowed their banks and formed vast inland lagoons. The flow of the water in the rivers being very slow, the tide, meeting the fresh water, helped to drive it back, and the mouths of the rivers gradually silting up, the waters became lakes for a great part of the year. In this way silt was deposited on the beds of these lakes and on this silt grew various aquatic plants. The land then rose somewhat or the waters receded, vegetable growth in-creased and even large trees grew, the trunks of which are still found in the peat. Again this land became submerged, a fresh deposit of silt occurred, and again plants, destined later to be turned into peat, flourished. In some places this must have occurred several times, as there are several layers of silt and turf.

The Romans appear to have made some attempt at embanking, but though the subject of draining the fens was considered in the fourteenth century, little seems to have been done for a century or so, till in 1490 John Morton, Bishop of Ely, had a river cut, 40 feet wide and

14 miles long, from Peterborough to Guyhirne. An Act was passed in the reign of Queen Elizabeth bearing on the subject, but nothing effectual was done till in 1630 a contract was made with Cornelius Vermuyden, a Dutch engineer, for draining the Great Level of the Fens.

There was so much opposition to this work being in

The Forty-foot or Vermuyden's Drain

the hands of a foreigner, that a request was made to Francis, Earl of Bedford, that he would undertake it. On January 13th, 1630–31, a contract to drain the fens was signed with the Earl, who, shortly after, obtained the co-operation of thirteen other gentlemen as joint "adventurers." The adventurers were to receive as payment 95,000 acres; of this they were to hand over

40,000 acres to maintain the drainage works in repair, and 12,000 to the King, while the remaining 43,000 were to be held by the Earl and his joint "adventurers."

In spite of the opposition that had been shown to Vermuyden, the adventurers employed him as their engineer. One of their principal undertakings was the cutting of what is now called the Old Bedford river, from Earith in the eastern corner of Hunts north-east to Salter's Lode, 70 feet wide and 21 miles long. After some years' work and an expenditure of £100,000 the drainage was declared inadequate by the Government at a session of Sewers held at St Ives, October 12th, 1637; and not until 1653, when William, Duke of Bedford, had succeeded his father and further work had been carried out at great expense, was the undertaking declared complete.

It will be seen that the principle adopted by Vermuyden was to make little use of the natural rivers, but to dig long, wide drains—the Old Bedford river is an instance of this—then narrower ones as feeders to them, and smaller drains still, feeding these, so that all were connected like a network. Windmills were then placed at various points to act as pumps to raise the water from one level to another. Lately these windmills have been superseded by steam-pumps. By this means the depth of the water is easily regulated and—though a disastrous inundation, owing to the breaking of Denver sluice, occurred in 1862, while 1879 witnessed the flooding of many acres of land—skill, perseverance, and the expenditure of a large sum of money have made of a district that was a morass in winter, and only capable of feeding cattle in summer, one of the most productive areas in England.

The Meres.

But this drainage of the fens, taking as it did only the water to a certain level, left several meres undrained. Whittlesey Mere, Trundle Mere, Ugg Mere, Ramsey

The Site of Whittlesey Mere

Mere, and Benwick Mere, all of these remained as meres until well into the nineteenth century. The Ramsey Chronicler, writing in monastic times, speaks of Ramsey Mere as being "a delightful object to beholders: in the deep and great gulfs of which Mere, there are frequently taken, by several sorts of nets, as also with baited hooks and other fishing instruments, pikes of an extraordinary bigness called hakedes by the country people: and though both fishers and fowlers cease neither day nor night to haunt it, yet there is always of fish and fowl no little store."

This lake and Ugg Mere were drained about 1840,

Whittlesey Mere.   Engine for draining the Mere

and in 1850 the work of draining Whittlesey Mere, the last and largest of the Fenland lakes, was completed, having been undertaken as a private enterprise by Mr Wells of Holme, to whom it belonged.   This piece of water is stated by Camden to have been six miles long, but on the map published by Mr Bodger in 1786 it is said to be three and a half miles from east to west and two and a half miles from north to south. The Mere was emptied in 1850, but, the banks giving way, was again filled. It was finally emptied in 1852, leaving an area of some 3000 acres to be changed from a peat-covered swamp into agricultural land.

It had been noticed previously, peat being very much like a sponge, that as the water was withdrawn the surface sank, so, in order to test the amount of this shrinkage, an iron post was driven into the gault, leaving the top on a level with the surface of the soil. In 1860 four feet nine inches of the post were exposed, in 1875 the surface had sunk eight feet two inches, and in 1909, the post registered a depression of nine feet ten inches.

5. Watershed. Rivers.

Huntingdonshire has two large rivers, the Nene and the Ouse. The main watershed of these rivers is in the higher lands of Northamptonshire, Oxfordshire, Buckinghamshire, and Bedfordshire, while some of their tributaries rise in the high ground on the west of the county itself.

The Nene rises in the western part of Northamptonshire and first reaches Huntingdonshire at Elton, from which place it forms the western and northern boundary of the county till it comes to Peterborough. At Stanground Sluice, a mile south-east of that city, the river divides into two branches, one going north-east and the other still forming the county boundary for another mile, till at Horsey Sluice it turns south-west, and takes a circuitous route through the north-eastern part of the county.

This branch, now an insignificant stream, was at one time the main river, its waters finding an outlet into the Wash by means of a junction with the Ouse at Salter's Lode. In its course it passed through the far-famed

The River Ouse at Hartford

Whittlesey Mere, and also through Ugg and Ramsey Meres, finally leaving the county near Benwick.

The Ouse rises near Brackley in Northamptonshire, passes through the counties of Northampton, Oxford, Buckingham, and Bedford, and enters Huntingdonshire near St Neots, flowing thence nearly due north to Huntingdon, where it turns and takes an easterly course to St Ives: then, a few miles further, it forms the boundary between Cambridgeshire and Huntingdonshire.

At Earith Sluice it divides into two rivers. One called the West Water, running north towards Chatteris ferry and joining the old course of the Nene near Benwick, is now nearly grown over and obliterated, though in many places it still forms the boundary between the Isle of Ely

The Ouse at the Paper Mill, St Neots

and Huntingdonshire. The other, called the Old West river, runs in an easterly direction towards Ely, three miles south of which city it receives the Cam or Granta; it then flows on past Denver Sluice and Salter's Lode, where it receives the combined waters of the old Nene and the West Water, and then flows direct to the Wash.

Numerous brooks and watercourses draining the

county feed these two rivers. Among others the Billing Brook runs into the Nene near Water Newton, and the Broughton Brook into the old course of the Nene below Ramsey.

The Ouse receives the Kym, the Alconbury Brook, the Ellington Brook, and other smaller tributaries.

6. Geology and Soil.

By Geology we mean the study of the rocks, and we must at the outset explain that the term *rock* is used by the geologist without any reference to the hardness or compactness of the material to which the name is applied; thus he speaks of loose sand as a rock equally with a hard substance like granite.

Rocks are of two kinds, (1) those laid down mostly under water, (2) those due to the action of fire.

The first kind may be compared to sheets of paper one over the other. These sheets are called *beds*, and such beds are usually formed of sand (often containing pebbles), mud or clay, and limestone, or mixtures of these materials. They are laid down as flat or nearly flat sheets, but may afterwards be tilted as the result of movement of the earth's crust, just as you may tilt sheets of paper, folding them into arches and troughs, by pressing them at either end. Again, we may find the tops of the folds so produced worn away as the result of the action of rivers, glaciers, and sea waves upon them, as you might cut off the tops of the folds of the paper with a pair of shears.

This has happened with the ancient beds forming parts of the earth's crust, and we therefore often find them tilted, with the upper parts removed.

The other kinds of rocks are known as igneous rocks, which have been melted under the action of heat and become solid on cooling. When in the molten state they have been poured out at the surface as the lava of volcanoes, or have been forced into other rocks and cooled in the cracks and other places of weakness. Much material is also thrown out of volcanoes as volcanic ash and dust, and is piled up on the sides of the volcano. Such ashy material may be arranged in beds, so that it partakes to some extent of the qualities of the two great rock groups.

The relations of such beds are of great importance to geologists, for by means of these beds we can classify the rocks according to age. If we take two sheets of paper, and lay one on the top of the other on a table, the upper one has been laid down after the other. Similarly with two beds, the upper is also the newer, and the newer will remain on the top after earth-movements, save in very exceptional cases which need not be regarded here, and for general purposes we may look upon any bed or set of beds resting on any other in our own country as being the newer bed or set.

The movements which affect beds may occur at different times. One set of beds may be laid down flat, then thrown into folds by movement, the tops of the beds worn off, and another set of beds laid down upon the worn surface of the older beds, the edges of which will abut against the oldest of the new set of flatly deposited beds, which latter may in turn undergo disturbance and renewal of their upper portions.

Again, after the formation of the beds many changes may occur in them. They may become hardened, pebble-beds being changed into conglomerates, sands into sand-stones, muds and clays into mudstones and shales, soft deposits of lime into limestone, and loose volcanic ashes into exceedingly hard rocks. They may also become cracked, and the cracks are often very regular, running in two directions at right angles one to the other. Such cracks are known as *joints*, and the joints are very important in affecting the physical geography of a district. Then, as the result of great pressure applied sideways, the rocks may be so changed that they can be split into thin slabs, which usually, though not necessarily, split along planes standing at high angles to the horizontal. Rocks affected in this way are known as *slates*.

If we could flatten out all the beds of England, and arrange them one over the other and bore a shaft through them, we should see them on the sides of the shaft, the newest appearing at the top and the oldest at the bottom, as in the annexed table. Such a shaft would have a depth of between 10,000 and 20,000 feet. The strata beds are divided into three great groups called Primary or Palaeozoic, Secondary or Mesozoic, and Tertiary or Cainozoic, and the lowest of the Primary rocks are the oldest rocks of Britain, which form as it were the foundation stones on which the other rocks rest. These may be spoken of as the Pre-Cambrian rocks. The three great groups are divided

	Names of Systems	Subdivisions	Characters of Rock
TERTIARY	**Recent Pleistocene**	Metal Age Deposits Neolithic ,, Palaeolithic ,, Glacial ,,	Superficial Deposits
	Pliocene	Cromer Series Weybourne Crag Chillesford and Norwich Crags Red and Walton Crags Coralline Crag	Sands chiefly
	Miocene	Absent from Britain	
	Eocene	Fluviomarine Beds of Hampshire Bagshot Beds London Clay Oldhaven Beds, Woolwich and Reading Thanet Sands [Groups	Clays and Sands chiefly
SECONDARY	**Cretaceous**	Chalk Upper Greensand and Gault Lower Greensand Weald Clay Hastings Sands	Chalk at top Sandstones, Mud and Clays below
	Jurassic	Purbeck Beds Portland Beds Kimmeridge Clay Corallian Beds Oxford Clay and Kellaways Rock Cornbrash Forest Marble Great Oolite with Stonesfield Slate Inferior Oolite Lias—Upper, Middle, and Lower	Shales, Sandstones and Oolitic Limestones
	Triassic	Rhaetic Keuper Marls Keuper Sandstone Upper Bunter Sandstone Bunter Pebble Beds Lower Bunter Sandstone	Red Sandstones and Marls, Gypsum and Salt
PRIMARY	**Permian**	Magnesian Limestone and Sandstone Marl Slate Lower Permian Sandstone	Red Sandstones and Magnesian Limestone
	Carboniferous	Coal Measures Millstone Grit Mountain Limestone Basal Carboniferous Rocks	Sandstones, Shales and Coals at top Sandstones in middle Limestone and Shales below
	Devonian	Upper } Mid } Devonian and Old Red Sand- Lower } stone	Red Sandstones, Shales, Slates and Lime- stones
	Silurian	Ludlow Beds Wenlock Beds Llandovery Beds	Sandstones, Shales and Thin Limestones
	Ordovician	Caradoc Beds Llandeilo Beds Arenig Beds	Shales, Slates, Sandstones and Thin Limestones
	Cambrian	Tremadoc Slates Lingula Flags Menevian Beds Harlech Grits and Llanberis Slates	Slates and Sandstones
	Pre-Cambrian	No definite classification yet made	Sandstones, Slates and Volcanic Rocks

into minor divisions known as systems. The names of these systems are arranged in order in the table, and the general characters of the rocks of each system are also stated.

With these preliminary remarks we may now proceed to a brief account of the geology of Huntingdonshire. Practically the whole of the county rests on the great bed of the Oxford Clay, which, however, is almost invariably covered by a considerable thickness of boulder clay and gravel, except in the north and east, where the covering is the alluvium of the Fens.

The Oxford Clay may be seen in the brickyards at St Neots, where it is covered by a bed of gravel; at the brickyards at Godmanchester, where the covering consists of some 18 feet of boulder clay; at St Ives, where a band of Corallian rock occurs; as well as in the brickyards of the Fletton district, Ramsey and the neighbourhood, and Warboys. Borings have been made at Abbots Ripton to a depth of 180 feet, and at Bluntisham to a depth of 300 feet, while that at Ramsey was carried down to 302 feet; but none of these reached the bottom of the Oxford Clay, which is estimated to be from 400 to 600 feet thick.

In the north-west part of the county, in the neighbourhood of Stibbington and Water Newton, we find an interesting section of the different beds lying one above the other, but underlying the Oxford Clay, though appearing just at this point near the surface of the ground.

Below the Oxford Clay comes first Cornbrash, then

Great Oolite Clay, followed by Great Oolite Limestone, Upper Estuarine Series, Lincolnshire Limestone, Lower Estuarine Series, and Northampton Sands and Ironstone, in the above order going downwards; all these rest upon a bed of Upper Lias Clay. Of these the Cornbrash is a hard blue limestone, the exposed surface of which tones to a brown colour, as may be seen in the valley of the Billing Brook; this stone is used for road-making. The Great Oolite beds which crop out at Alwalton Lynch take the form of a shelly limestone, which has been named Alwalton Marble, and was much used in olden days in the neighbouring churches, and very largely in Peterborough Cathedral. The Upper Estuarine Series, which consists of sands and clays varying in thickness from three to nine feet with nodular ironstones towards their base, may be seen in the brickyards at Water Newton.

Below this series lies the Lincolnshire Limestone—a fine-grained Oolite stone very largely quarried in the neighbouring county of Lincoln, where it assumes the character of a valuable building stone. It only just extends into Huntingdonshire and is exposed in shallow beds at Sibson tunnel.

Below this formation are to be found the Northamptonshire Sands and Ironstones, which at this point are some 16 or 18 feet thick.

Above the Oxford Clay usually lies, as already stated, a thick bed of boulder clay. This is of a bluish colour and contains boulders and large fragments of rocks of various kinds, flint, and, in some places, beds of gravel; occasionally, as at Great Catworth, it also contains chalk.

All the "highland" of the county—as the hilly, undulating district occupying its southern and western part is called—is formed of this boulder clay, the surface of which is well adapted for cultivation and is largely utilised for growing wheat and other cereals; some of this highland is however devoted to pasturage.

At Waresley and Great Gransden occur beds of Lower Greensand. The river valleys—the soil of which consists of drift, comprising clay, gravel, sands, fossils, etc.—are in many places of considerable breadth, forming rich meadows. Deposits of gravel are found at Hemingford, Hartford, Bluntisham, and Somersham.

The northern and eastern parts of the county, belonging to the great district known as "The Fens," are covered by an alluvial deposit of marine silt and peat; and, as we have seen, very frequently two, and sometimes more, beds or layers of each are found of varying thicknesses. It is clear that the lower silt bed was deposited on the clay at a time when this district was under water, then on this silt bed grew various mosses and other plants; until in some way, either by a sinking of the surface or by the silting up of the mouths of the rivers, or both, the whole of the district was again submerged, and a fresh deposit of silt was made above the mosses. Then again this bed rose above the water, either by some convulsion of nature or the receding of the water, and once again mosses and plants took root and grew upon this second layer of deposit. On this grew trees of considerable size, and many trunks of oaks and birch are met with by the fen drainers, as well as the remains of the wild ox, red deer, bear, beaver, and wolf—now long ago extinct in their wild state—together with those of animals still to be found at the present day, as for example, the otter and the fox.

7. Natural History.

The fauna and flora of a country or district depend upon a great variety of factors. So numerous, indeed, are these, that we need not attempt to give more than a rough outline of them here. Moreover, they are in a great degree interdependent. In the distribution of species, whether of plants or animals, the character of the soil may play a considerable part, the dampness or dryness of the atmosphere, elevation above sea level, coldness or warmth, and so forth. But it will be seen that these all react upon each other, and on the whole it may be said that, with regard to animals, the presence or absence of suitable food is the chief cause influencing their distribution, no less in a small district than in a great continent. At the same time we cannot explain everything by this, and certain birds—as the starling, for example—have extended or altered their range in a very remarkable manner with no very apparent reason.

In Great Britain—still taking birds as our illustration as the most familiar objects—we are accustomed to associate certain parts of the country with certain kinds of birds. We do not look, for example, for the birds of the mountain and the moor in a Surrey cornfield. The oozy flats of tidal creeks, again, are haunted by quite another class, while it is not surprising to learn that our southern seaboard receives the greatest number of the rarer visitors from the continent and that Cornwall is richest in American species. More than any other, perhaps, the fen country has its own characteristic belongings. It swarmed with waterfowl in old days and

Brampton Mill

fish were plentiful, so that with an abundant and easily obtainable food-supply there was no lack of people ready to brave the fevers of the marshes and the rigours of their bitter winters.

If we look at the map of the county we shall see that while none of its surface is of any great elevation, the whole of the north-east, indeed about a quarter of the entire county, is true fenland. Here were the four meres, now no longer in existence—Ramsey, Whittlesey, Trundle, and Ugg—and here the characteristic fauna and flora pertaining to this part of our land are still apparent, though all the meres are dry and there is but little undrained and unreclaimed land except a part of Walton Fen. Of the birds a great number of marsh-haunting species still remain, but there are many, once common, which have either completely disappeared or are now to be recorded only as rarities. Such are the bittern and the little bittern, and the grey-lag goose, which nested freely in the fenland of the eastern counties a hundred years ago. Doubtless the ruffs and reeves were plentiful in Huntingdonshire too, as in Lincolnshire and Norfolk, but they are now gone, while in ancient days the pelican must have lived here, for his bones have been found. Two birds of exceptional interest must be mentioned which in quite recent times have passed out of the list of our avifauna never to return. The first of these is the so-called bearded tit (*Panurus biarmicus*), one of the most beautiful birds in England, still to be found in places in the Norfolk Broads in diminishing numbers, and once common both in Whittlesey and Ramsey Meres. The other is Savi's warbler (*Locustella luscinioides*), which, only recognised as a distinct species in 1824, though probably abundant, was drained out of the fenland by the middle of the last century.

But if we have lost these, we have still left others characteristic enough—the sedge warbler, mimic of half-a-dozen other birds and a delightful songster of the night, the reed warbler, another night singer, and the grasshopper warbler, whose curious whirring note has given it the name of the reeler among the fenmen. Here, too, is the handsome reed bunting, and the common sandpiper, and

The Large Copper Butterfly (*Polyommatus dispar*)

most of the marsh-loving kinds. But mammals, as might be imagined, are not very common in this district, the otter and water-vole excepted. The polecat (*Mustela putorius*), is, however, said to be met with still.

More noteworthy are the butterflies of the fens. The large copper (*Polyommatus dispar*) has now been extinct for more than half a century. Though isolated instances of its occurrence have been claimed for Cambridgeshire, there are few records of them, and the reputed captures in Norfolk and Suffolk of old times are all dubious. Huntingdonshire may justifiably claim it as peculiar to the county, at least in historic times, the fens round Whittlesey, Ramsey, and Ugg being its chief haunt, and Holme Fen its last stronghold. The beautiful swallow-tail (*Papilio machaon*), which still lingers in Wicken Fen in Cambridgeshire, was common in the districts mentioned above, but it has now disappeared altogether and has not been seen in our county since 1870. The drainage of the fens caused the disappearance of species not only by the destruction of the plants forming the food of the larvae, but, in the case of the swallow-tail, by the sedge-cutters taking away the sedge to which the chrysalides were attached. The purple emperor (*Apatura iris*), a very fine species, inhabits Monks Wood. But here we are passing from the subject of the fens.

Let us turn to the plants. Cultivation has altered the face of the land, and great stretches of corn and potatoes have taken the place of the fen plants of former days and brought with them the common annual weeds we see associated with them in the dry lands. Then, too, with the making and repairing of roads with material from elsewhere many other roadside plants have been introduced which are not really native to the district. The flatness of the country strikes the eye of a stranger at once : it is rendered more conspicuous, no doubt, by the fact that ditches serve the function of hedges. Here flourish the reeds, and here in spring the marsh marigold makes the ground gay with its golden cups. The sedge (*Cladium mariscus*) is a true fen plant, and is used for thatch, or, mixed with grass and other plants, for bedding for cattle. Other characteristic plants of the fen are the marsh fern (*Lastraea palustris*), the orange-flowering *Thalictrum flavum*, the marsh pea (*Lathyrus palustris*), the water-lily, the water-dock, and the familiar rush and bulrush. Less common, and needing more search owing to their modest habit, are the sundew and the strange bladderwort (*Utricularia*).

Kingsley, in his *Prose Idylls*, writing of Whittlesey Mere in bygone days, says :—"But grand enough it was, that black ugly place, backed by Caistor Hengland and Holme Wood and the patches of primaeval forest, while dark green alders and pale green reeds stretched for miles round the broad lagoon, where the coot clanked and the bittern boomed, and the sedge-bird, not content with its own sweet song, mocked the notes of all the birds around; while high overhead hung motionless hawk beyond hawk, buzzard beyond buzzard, kite beyond kite, as far as the eye could see. Far off, upon the silver mere, would rise a puff of smoke from a punt, invisible from its flatness and its white paint. Then down the wind came the boom of the great stanchion gun, and after that sound another sound, louder as it neared, a cry as of all the bells of Cambridge and all the hounds of Cottesmore, and overhead rushed and whirled the skeins of terrified wildfowl, screaming, piping, clacking, croaking, filling the air with the hoarse rattle of their wings, while clear above all sounded the wild whistle of the curlew and the trumpet of the great wild swan. They are all gone now. No longer do the ruffs trample the sedge into a hard floor in their fighting rings, while the sober reeves stand round admiring the tournament of their lovers, gay with ears

A Woodland Road, St Neots

and tippets, no two of them alike. Gone are ruffs and reeves, spoonbills, bitterns, avocets: the very snipe, one hears, disdain to breed."

As has been stated, the southern and western parts of Huntingdonshire show a very different sort of country. Here we have a slightly undulating upland with pleasant scenery, though devoid of any striking feature and deficient in woodland. This district has a flora hardly differing from that of the neighbouring counties. In spite of the lack of trees, oaks grow to a large size in some places, and in Elton Park are some very fine though decaying specimens which are believed to date back nearly to the Conquest. Within our county, too, originated the Huntingdonshire elm (*Ulmus glabra*) about a century ago.

8. Climate.

The climate of a country or district is, briefly, the average weather of that country or district, and it depends upon various factors, all mutually interacting—upon the latitude, the temperature, the direction and strength of the winds, the rainfall, the character of the soil, and the proximity of the district to the sea.

The differences in the climates of the world depend mainly upon latitude, but a scarcely less important factor is this proximity to the sea. Along any great climatic zone there will be found variations in proportion to this proximity, the extremes being "continental" climates in the centres of continents far from the oceans, and "insular" climates in small tracts surrounded by sea. Continental climates show great differences in seasonal temperatures, the winters tending to be unusually cold and the summers unusually warm, while the climate of insular tracts is characterised by equableness and also by greater dampness. Great Britain possesses, by reason of its position, a temperate insular climate, but its average annual temperature is much higher than could be expected from its latitude. The prevalent south-westerly winds cause a drift of the surface-waters of the Atlantic towards our shores, and this warm-water current, which we know as the Gulf Stream, is the chief cause of the mildness of our winters.

Most of our weather comes to us from the Atlantic. It would be impossible here within the limits of a short chapter to discuss fully the causes which affect or control weather changes. It must suffice to say that the conditions are in the main either cyclonic or anticyclonic, which terms may be best explained, perhaps, by comparing the air currents to a stream of water. In a stream a chain of eddies may often be seen fringing the more steadily-moving central water. Regarding the general north-easterly moving air from the Atlantic as such a stream, a chain of eddies may be developed in a belt parallel with its general direction. This belt of eddies, or cyclones as they are termed, tends to shift its position, sometimes passing over our islands, sometimes to the north or south of them, and it is to this shifting that most of our weather changes are due. Cyclonic conditions are associated with a greater or less amount of atmospheric disturbance ; anticyclonic with calms.

The prevalent Atlantic winds largely affect our island in another way, namely in its rainfall. The air, heavily laden with moisture from its passage over the ocean, meets with elevated land-tracts directly it reaches our shores—the moorland of Devon and Cornwall, the Welsh mountains, or the fells of Cumberland and Westmorland —and blowing up the rising land-surface, parts with this

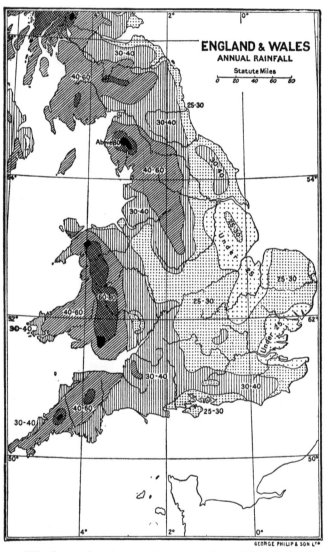

ENGLAND & WALES
ANNUAL RAINFALL
Statute Miles
0   20   40   60   80

30-40

40-60

25-30

30-40

Above 80

40-60

30-40

Under 25

60-80

40-60

30-40

25-30

25-30

Under 25

30-40

40-60

30-40

30-40

25-30

GEORGE PHILIP & SON Lᵗᵈ

(The figures give the approximate annual rainfall in inches.)

moisture as rain. To how great an extent this occurs is best seen by reference to the accompanying map of the annual rainfall of England, where it will at once be noticed that the heaviest fall is in the west, and that it decreases with remarkable regularity until the least fall is reached on our eastern shores. Thus in 1906, the maximum rainfall for the year occurred at Glaslyn in the Snowdon district, where 205 inches of rain fell; and the lowest was at Boyton in Suffolk, with a record of just under 20 inches. These western highlands, therefore, may not inaptly be compared to an umbrella, sheltering the country further eastward from the rain.

The above causes, then, are those mainly concerned in influencing the weather, but there are other and more local factors which often affect greatly the climate of a place, such, for example, as configuration, position, and soil. The shelter of a range of hills, a southern aspect, a sandy soil, will thus produce conditions which may differ greatly from those of a place—perhaps at no great distance—situated on a wind-swept northern slope with a cold clay soil. In Huntingdonshire we have a large low-lying area backed up upon the south and west by higher land, the prevailing slope of which is towards the north, and the soil itself in the fens is of a peaty and marsh nature, and on the highlands chiefly clay, all which tend to promote dampness of the atmosphere.

The average rainfall in Huntingdonshire in 1908 was 19·85 inches, whereas at The Stye amidst the hills of Cumberland the rainfall was 177·25 inches, at Princetown in South Devon 69·50, and at Swansea in Wales 135·08, while on Snowdon as much as 237·32 inches were registered.

The prevailing winds of Huntingdonshire are from the west.

There are very few stations in Huntingdonshire for recording the number of hours of sunshine, but the average for the whole of England in 1907 was 1560 hours during the year, and that of Cambridge 1578 hours, and the two counties being contiguous and each having a large area of fenland, it is probable that the number of hours of sunshine is rather above than below the average.

9. People—Race, Dialect, Population.

We know little about the ancient people who made and used the flint implements which are found in the river gravels around Somersham and at other places, and even when we come to the latter part of the Stone Age, though we find skeletons in the barrows or mounds upon our downs, our information about the race is exceedingly small. This is perhaps not to be wondered at, for in no case do we find any knowledge of the art of writing in the stage of culture when only stone and no metal implements were used. Moreover, we must bear in mind that all we know about early England from written history is from the works of foreign merchants or of foreign conquerors.

It is a natural law that primitive races, when brought into contact with stronger and conquering peoples, either become absorbed by them or take refuge in the mountains or least accessible parts of the country. No Welsh hills offered a refuge for the dweller in our county, but the Fenland was a scarcely less sure haven, and here, accordingly, we might look for instances of the original types, if any such should chance to exist. It would seem that they do, for here we occasionally find men whose characteristics mark them out as being totally distinct from the bulk of their neighbours, and these may well be the descendants of Britons driven to the fastnesses of the fens as most of their fellow-tribesmen were driven westward.

The Huntingdonshire inhabitants of to-day, however, are mostly sprung from Teutonic stock, chiefly Angles, slightly intermixed with some Danish blood; and as the Angles or Engles had force enough to give their name to the whole country, and have called it England, so their language has become the standard language of all English peoples. The authors of *Fenland, Past and Present*, Messrs Miller and Skertchley, hold that "the Fenland was the cradle of modern classic English; that here was the fusion of those elements of speech into a dialect which was to grow into a model language, a vehicle of thought for a great nation, a language more widely used in the present day than any other on the globe."

It is a noteworthy fact that the place-names of Huntingdonshire are wholly of Saxon, that is Angle, origin; there are only three Danish names in the whole county, viz. Upthorpe, Sibthorpe, and Daintree, and these are not the names of villages, but simply of hamlets which originally were no doubt merely the residences of individual Danish settlers. And this Saxon nomenclature is especially emphasised in the name transformation of the Roman word Castra. Wherever this occurs in Huntingdonshire, it invariably appears as *Chester*, thus Godmanchester, Chesterton; whereas in the more Danish counties of Lincoln and Northamptonshire it is *Caster*, as Casterton, Ancaster, etc.

The typical Huntingdonshire man of to-day is rather above the medium height, with brown hair and dark eyes and regular features, not differing much from the race to be found in many parts of middle England, but with a tendency nevertheless to the more handsome and finer rather than to the coarse and heavy types found in some parts. The women are noted for their clear complexions and well-cut features.

In the seventeenth century the drainage of the Fens brought in many Dutchmen and Walloons. At Thorney in Cambridgeshire, on land belonging to the Duke of Bedford, a strong colony was formed of these people, who had a church with minister and registers of their own ; and from these foreigners, and to some extent from Huguenot refugees, are derived many families now settled in this and the adjoining counties, for instance those of Le Plas, Descou, Le Fevre, Vergette. A few also of these foreign names are derived from the French prisoners interned at Norman Cross at the beginning of the nineteenth century, some of whom settled in the neighbourhood after the conclusion of peace.

On examining the returns made at the time of the Domesday survey, and taking the average family to consist of five persons, we may conclude that the population of Huntingdonshire was then about 12,500. But it must be remembered that at that time much of the upland portion of the county was forest or waste land, while other great portions consisted of fen and meres.

When the first systematic census was taken in 1801, the population was 37,568, and this had increased to 53,192 by 1831, and to 64,250 (its highest point) in 1861. Since then each succeeding census has shown a steady decline until 1891, when it stood at 57,761.

In 1901, the population of the ancient or geographical county numbered 57,771, or just 10 more than at the previous census, and rather more than four times the number residing here at the time of Domesday.

It must be remembered, however, that many of the villages have a smaller population to-day than they had in 1831; the allurements of town life, and the intro-duction of labour-saving machinery rendering less manual labour necessary on the land, have both contributed to the depletion of our village population. But on the other hand there has been a large increase in the population of Fletton, Woodston, Stanground, and Yaxley, on account of the brickfields that have been opened at those places; while many of the towns have had a substantial increase, Ramsey increasing from 3006 in 1831 to 4823 in 1901, and Huntingdon from 3267 to 4349. St Neots shows a slight increase, but St Ives and Godmanchester have both diminished somewhat, and Kimbolton, partly owing to the decline in the lace trade, and partly to the fact that it is rather far from a main line of railway, has decreased from 1584 to 915.

It is interesting to notice that of the 57,771 inhabitants, 38,955 were born in the county, while 4601 were born in the neighbouring county of Cambridge and 3250 in Northamptonshire. Foreigners are very few in number, there being only 37 in the whole county, and of these France heads the list with eight, then follow the United States and Germany with six each, Italy with five, Russia and Switzerland with three, and Norway, Denmark, Holland, Belgium, Spain, and some country in Africa unnamed have one each.

Huntingdonshire being a purely agricultural county, it is naturally rather thinly populated. The average number of people to the square mile in England and Wales is 558, but in Huntingdonshire it is only 157, or about one inhabitant to every four acres, whereas in Lancashire the population is fifteen times as thick.

The proportion that agriculture bears to other industries may be gauged by comparing the figures given in the census returns. Thus, of the 26,735 males over 16 years of age in this county, no less than 7596, or nearly one-third, are engaged in agriculture, while those occupied in brickmaking and kindred trades number 1210, and 713 are employed on railways.

Agriculture was ever a healthy occupation, and consequently many of the inhabitants of Huntingdonshire attain a venerable old age. In 1901, of the total population of the county, 1951 males and 2231 females were over 65 years of age, but there were no persons over the age of 100 years.

10. Agriculture — Main Cultivations. Woodland. Stock. Treatment of Fen Soil.

On consulting the diagrams dealing with agriculture given at the end of this book, we see that the area of the county amounts to 234,218 acres, and that of this total 86,522 acres form permanent pasture. A moderate acreage only produces clover and rotation grasses. We have thus nearly 140,000 acres left, and almost the whole of this is arable land, in other words nearly two-thirds of the county. Huntingdon is thus conspicuously an agricultural county.

About one-fifth of this area is occupied by wheat ; the stiff clay lands of some parts of the county being specially suitable for growing this cereal. Barley, oats, and beans are also largely grown, but only a small area is sown with rye. In addition to the above, potatoes, turnips, and mangolds are much cultivated, and some parts of the highland are used for the growth of clover and kindred crops for seed and fodder. Here and there one sees a crop of flax, and near Fenstanton a rather unusual plant—chicory—is grown.

The district about Somersham and Bluntisham, in the south-east of the county, is the chief neighbourhood for small-fruit cultivation, and here gooseberries, currants, strawberries, plums, and apples are produced in considerable quantities. Bluntisham, Earith, Somersham, Colne, Hemingford Grey, and Fenstanton have between them an area of over 600 acres under fruit.

The area occupied by woods is 5158 acres. In them we see the oak, ash, elm, birch, and beech, with an undergrowth of hazel and other bushes.

Of the domestic animals in Huntingdonshire, sheep are more numerous than all other farm-stock put together,

Shire Horse "Prince William"
(*The property of J. Rowell, Esq. of Bury, Hunts.*)

numbering 91,040. Cows and other cattle number 29,387, pigs 19,444, and horses 11,620.

Huntingdonshire and the neighbouring county of Cambridgeshire may be said to be the home of the

"Shire" horse. Those used in the county for farm purposes are principally of this breed.

The rich meadows of the south-west of the county afford good grazing for cattle, and indeed the county generally feeds a considerable number of stock for market every year. The cattle are mostly of the short-horn breed.

Potato-lifting in the Fens (new method)

A breed of pigs raised at Holywell in this county is known all the world over, and many are shipped to America and other countries.

In the Fens are grown in alternating rotation crops of wheat, barley, and oats; mangolds, turnips, and kohl rabi are also grown here, as well as buck-wheat, seed grasses, mustard, and in some parts rye. Potatoes from this district may be seen in London under the name of

4—2

"Blacklanders," and celery from the Fens is well known in Covent Garden market. In fact, potatoes and celery constitute the staple crops of the Fen district. Essex, which is more than four times the area of Huntingdonshire, has only 4648 acres in potatoes against 8459 in Hunts.

A great change has taken place in the agriculture of the county in the last century or so, and in the value of the Fenland, which has been rendered much more productive by the drainage system. Excellent crops are now grown on land which a century ago produced scarcely anything, and the area of cultivated land has been largely increased.

Some idea of the magnitude of this change may be formed by a comparison of the above account with a report on the agriculture of the county of Huntingdon by Mr George Maxwell of Fletton, published in 1793.

He says that the Fen lands yield but little profit, and states the area to be 44,000 acres—three-quarters of this land producing not more than 1/6 or 1/- an acre rent, and some none at all, whereas now a rent of 60/- per acre is not uncommon. He gives the area of the county as 240,000 acres, of which 130,000 are commonable land, 23,000 enclosed arable and woodland, 43,000 enclosed pasture, and 44,000 pasture.

The common field of a village was divided into three parts, one of which was annually fallowed, one part was sown with wheat followed by oats, the other part with barley followed by peas and beans. Thus every part produced the same crop every six years. At the present time

no commonable fields remain in the county, the whole of
them having been enclosed in the eighteenth and nineteenth centuries.

Mr R. Parkinson writing in 1811 says that cabbages,
carrots, and potatoes were grown only for the table ;
mustard, flax, and hemp were amongst the crops then
produced.

A Fen Farm near Ramsey Mere

The peat of the Fen being very light, it requires
special treatment to prevent the soil blowing away and to
make a good seed bed which the growing plants can
grasp. Below the peat lies clay or silt, and the practice
is to dig trenches from end to end of the fields a few yards
apart, and throw up the clay, covering the whole of the

land; this is then ploughed in, and a good surface is formed. This process is called gaulting.

Where gaulting has been neglected clouds of dust may sometimes be seen during very high winds, ditches become choked with the soil, and the young plants are sometimes removed from their position—crops of oats have thus been entirely carried away. In some parts where the gault is a long way below the surface, land is very difficult to cultivate, and lies practically waste, growing rough grass, alders, and the like. The passenger on the Great Northern Railway may see an instance of this on the east side of the railway just north of Holme station, where the iron post already alluded to, driven into the clay to ascertain the shrinkage of the peat, may likewise be seen. The surface continues to sink, owing to the drawing off of the water and the removal of the "turf" as the peat is termed, and gradually the surface will be lowered till it is possible to reach the clay. This can be done when the peat is not more than about six feet deep, and then more of this land will be brought into cultivation.

11. Industries and Manufactures.

The absence of coal and iron, the former not being found at all, and the latter occurring only in very small quantities, prevents Huntingdonshire from being a manufacturing county except on a small scale, and in products for the making of which the county possesses special facilities.

Bricks have been made in most parts of the county for very many years, but the smaller brickfields dispersed about the county have mostly ceased to be profitable in recent years, their place being taken by the extensive brickyards in the neighbourhood of Fletton, Farcet, and Yaxley, where very large quantities of bricks are made by machinery and sent to London and elsewhere.

Godmanchester Mill

There are extensive breweries at Huntingdon, St Neots, and St Ives, the water being considered particularly adapted for this industry. There are also large flour-mills at Godmanchester, St Neots, St Ives, Houghton, Offord, and Ramsey: and the county still has many of the picturesque windmills which have stood for centuries and are still in use for grinding corn for cattle and other farm stock.

At the paper-mills near St Neots William Fourdrinier erected in 1807 a machine capable of successfully manu-facturing paper in the continuous web, instead of in small

Houghton Mill, near St Ives

moulds, and this mill therefore may justly be called the cradle of the modern paper-making industry. It is still working to this day.

Lace-making, once very largely practised by the

peasant women of Huntingdonshire, after being almost
discontinued, has been considerably revived of late years.

At Godmanchester and Spaldwick, tanning was a
flourishing industry down to the early part of the nineteenth century, but has now ceased to exist.

Paper Mills, St Neots

Turf-cutting is still carried on, Walton Fen supplying
many thousand turves every year. The peat is cut in
brick-like blocks, stacked for a while, and when sufficiently
dry for fuel, is brought up to the towns and villages for
sale. A lighted turf, covered with ashes, left in the hearth
will burn for many hours, and the early-rising fenman has

a glowing fire in a few minutes when he has thus provided overnight for his requirements in the morning.

It should be mentioned that peat is almost an unknown word in Huntingdonshire, the word turf being always used. The rough, partly-decayed vegetable matter which is in process of becoming turf and lies just above it, is called "clunch," and that below the highest bed of silt

Spaldwick Village

or clay is called "moor." Clunch, turf, and moor are all used as fuel. Lately turf has been compressed into small blocks called "briquettes" with a view to extending the use of peat as fuel and so creating a large local industry, but the venture did not prove financially successful and has now been abandoned.

12. Fisheries. Fowling.

The last century has seen a great change in the fish supply for the inland counties of England.   Before the age of steam, very little fish from the sea, except such as had been salted, found its way far inland on account of the difficulty of transit.

All round our coasts the fishing industry is pursued, but especially on the east coast, where, the sea being shallow, large quantities of fish are taken.   At the present time several fleets of fishing vessels are engaged in this industry, each fleet fishes over a certain area, and steamers go round from vessel to vessel, and convey the fish rapidly to port : a constant service of these carrier steamers being kept up.   This arrangement enables the boats to remain at sea for weeks at a time, instead of returning each with its own "catch."   Grimsby is the great haven of the fishing fleets for this part of England, and from there Huntingdonshire receives the principal part of its fish supply.

In the old days inland fisheries were of much greater importance than they are at present.   Fish were used largely as food, and their capture employed a considerable number of men.   The Domesday account of the inhabitants of Huntingdonshire and their employments tells us that of a total of 2511 persons owning property, or living in the county, as many as 33 were fishermen, while at the same time there were only 33 handicraftsmen.   To-day few people, if any, in the county are simply fishermen, while the handicraftsmen have very much increased in numbers.

Eels were very favourite fish, the monks of Ramsey paying those of Peterborough 4000 eels a year for permission to quarry stone at Barnack; and ancient fishponds are still to be found close to the abbey, and also at Worlick, a mile and a half distant.

We find mention of pike (hakedes, or waterwolves as they are sometimes called in early deeds), and besides there were, and are, tench, carp, bream, roach, dace, perch, and gudgeon.

Whittlesey Mere, by far the largest and most important lake in the county, was granted in A.D. 864 by Wulphere, King of Mercia, to his new-founded monastery of Medehamstede, now called Peterborough; but at the time of the Domesday survey the Abbot of Ramsey had one boat on it, the Abbot of Peterborough one, and the Abbot of Thorney two. Of these two the Abbot of Peterborough held one, also two rights of fishery and two fishermen. We learn that "the Fisheries of the Mere of the Abbot of Ramsey bring him in ten pounds; those of the Abbot of Thorney, sixty shillings; and those of the Abbot of Peterborough, four pounds."

In 1614 there were 15 boat-gates belonging to Whittlesey Mere; these carried with them certain fishing rights, and as the names of the owners are carefully set out, it is evident these rights were of value. "To each boat gate did belong 40 pollenets, 40 wernets, 24 widenets, 24 bownets, 1 drave, 1 tramaile, also setting-lawe and syrelepes at the will of the owner."

The lord of the manor of Glatton and Holme, which included Whittlesey Mere, had a right to summon the fishermen of his manor to his two courts at Holme, and to prove their nets with a brazen "gougle," and he was allowed to take or destroy any nets with too small a mesh.

A Duck Decoy

There was also at Southoe in the southern part of the county a fishery producing 1000 eels a year at the time of the Domesday survey. Up to the end of the eighteenth century testators not infrequently describe themselves as fishermen or "fishers."

The fisherman was also sometimes a fowler, and even within the last 70 years fowling was a valuable industry.

Wild ducks, teal, and wigeon were caught by means of "decoys," which consisted of a pond, situated in the middle of an area of uncultivated land several acres in extent, where alders, sedge, and reed were allowed to grow. From the pond ran several curved ditches, some three or four yards wide where they left the pond and gradually narrowing to the other extremity which was at a distance of some 50 yards; at the end of each ditch was a net. In the pond swam tame ducks to act as decoys to their wild congeners. Ducks are inquisitive creatures, and when they were once in the pond would be incited by a well-trained dog to enter one of these drains; the dog, showing himself at intervals through gaps left in a screen along the side of the ditch, would draw the ducks on till they approached the net, when the fowler showing himself at the other end of the drain, many of the ducks flew straight into the net and were captured.

Mr Heathcote, in *Reminiscences of Fen and Mere*, tells us that as many as 200 dozen have been caught in seven days in a decoy still existing, though now disused, in Holme Fen.

13. History of Huntingdonshire.

When the Romans came to Britain they found it occupied by various tribes of the Celtic race, of which a powerful tribe called the Iceni inhabited the district now comprised by the counties of Norfolk, Suffolk, Cambridgeshire, and Huntingdonshire.

The Romans at first made an alliance with this tribe, but, at their later invasion in A.D. 43, they attacked and subjugated it after a very stubborn and gallant resistance. In A.D. 61, however, the Iceni revolted under their Queen Boadicea and were at first successful, but the Roman general, Suetonius Paulinus, returning from a victorious expedition against the Isle of Anglesey, met and utterly defeated them, and Boadicea is said to have died by her own hand. After this the whole territory of the Iceni was incorporated with the Roman province of Flavia Caesariensis.

When the Romans left Britain about A.D. 420, the inhabitants, finding themselves unable to resist the attacks of the Picts and Scots, called to their aid the Saxons, Jutes, and Angles or Engles, some of whom had probably already held a footing upon the eastern coast for many years. Of these tribes the Angles seem to have settled first in the district called from them "East Anglia," and later to have spread themselves over the north-east and central portions of England comprising the later kingdoms of Northumbria and Mercia. They seem to have come in constantly invading swarms, for Henry of Huntingdon tells us that as late as A.D. 527, "large bodies of men came successively from Germany and took possession of East Anglia and Mercia; they were not as yet reduced under the government of one king; various chiefs contended for the different districts, waging continual war with each other."

The kingdom of Northumbria was founded about 547; that of East Anglia a little later, Uffa being the first king;

while the kingdom of Mercia had its origin about the year 586.

In 571 a battle was fought between the army of the King of Wessex and the Britons at Bedford, in which the former were victorious, and thereupon overran Oxfordshire, Buckinghamshire, and Bedfordshire, but when they came to the place where the little river Kim runs into the Ouse, where the town of St Neots now stands, they went no further, for the reason, as may be surmised, that the Angles were already in possession of the opposite banks, and to this day the men of Huntingdonshire and Northamptonshire speak an Angle, and those of Bedfordshire a Saxon dialect.

East Anglia was converted to Christianity by Bishop Felix in the early years of the seventh century. After the death of Ethelbert, King of East Anglia in 792, Huntingdonshire was annexed to the kingdom of Mercia, and so continued until the Heptarchy came to an end in 827, when Egbert, King of Wessex, subjugated the other kingdoms but allowed them to elect their own kings, who held their countries as tributary to him. Shortly after this Northmen from Norway, Sweden, and Denmark commenced to ravage the coast and finally spread themselves over the whole country. Of these invaders, the Danes obtained complete possession of the eastern counties, whence they harried the neighbouring lands, until Alfred the Great, after his victory at Ethandune, in A.D. 878, concluded a treaty at Wedmore with the Danish leader Guthrum, whereby the latter was made governor of the northern and eastern part of England, a division which is said to have included that part of Huntingdonshire south of the Ouse. The settlement effected by the treaty did not, however, last very long. After the death of Guthrum in 890, and of Alfred in 901, numerous conflicts between the two nations, the Saxons and the Danes, took place, which was probably due to the fact that fresh bodies of Danes were still flocking to our shores. We find it

Toseland Manor House

recorded that in 921, the Danes, who had long occupied Huntingdon, forsook their fortress there and made another at Tempsford on the Ouse, in Bedfordshire, and began to harass the country towards Bedford. But King Edward the Elder having recovered Buckingham from them in 915, and Bedford in 919, advanced in 921 against this new outpost at Tempsford, where he slew a Danish

Jarl named Toli, who had seized Huntingdonshire, and whose name still survives in the name of the village and hundred of Toseland (Tolis-land). Edward at this time repaired and rebuilt the town of Huntingdon.

For nearly a hundred years the Saxon kings held almost undisputed sway, and the country settled down to peaceful pursuits; during which period many of the monasteries were founded, including the great Abbey of Ramsey (founded in 969), which afterwards owned possessions in nearly every parish in Huntingdonshire. But during the reign of Ethelred the Unready the Danes again began to gain the ascendency; in the year 1010 they landed and fought their way up the Ouse as far as Bedford, and devastated and burned the country as they went.

Eventually, in 1016, Canute became King of all England. He had a hunting-box at Bodsey, close to Ramsey, and gave his name to a "dike," or road with a ditch on each side, stretching from Ramsey towards Peterborough, which he caused to be made.

The years from the death of Canute, in 1035, until the Norman Conquest in 1066, were largely occupied by disputes between rival factions in the state, of which by far the most important was that between Godwin, Earl of Wessex, and his son Harold, Earl of East Anglia and Essex, on the one side, and Leofric, Earl of Mercia, and Siward, Earl of Northumbria, on the other.

Godwin was the chief leader of the Saxon party as opposed to the band of Norman retainers with whom King Edward the Confessor had filled his court; and these latter, aided by Leofric and Siward, obtained the banishment of Godwin and his sons in 1051, upon which Huntingdonshire and Cambridgeshire were separated from Mercia and added to East Anglia under the Earldom of Ælfgar, son of Leofric.   Godwin and Harold, however, returned the next year, and obtained restitution of their honours; but Godwin died in 1053, and Harold succeeded him as Earl of Wessex, whereupon Ælfgar again became Earl of East Anglia.

In 1057 the latter succeeded his father as Earl of Mercia, and Gurth, Harold's youngest brother, was made Earl of East Anglia.

Siward died in 1055, and Tostig, another brother of Harold, was made Earl of Northumbria, to which the counties of Northampton and Huntingdon were now added; but ten years later he was deposed, and Morcar, the younger son of Ælfgar, was made Earl in his stead; Northamptonshire and Huntingdonshire, however, being now formed into a separate earldom under Waltheof, the son of Siward.

It is important to remember these changes, for they explain the reason for many of the events which happened during this period.   For instance, about the year 1065 a combination of Northumbrians, Mercians, and Welshmen (strangers), under Edwin and Morcar, sons of Ælfgar, attacked the counties of Northampton and Huntingdon, then under Tostig, son of Godwin, killing, burning, and committing serious depredations, the effects of which were felt for years afterwards.

And again we see this same party-feeling showing

itself when Harold made his splendid march southward
from York to meet William the Norman on the field of
Hastings, for Edwin and Morcar refused assistance, but
the men of Northamptonshire and Huntingdonshire joined
his army, and Aluric, Sheriff of Huntingdonshire, is recorded as being amongst the slain. But it would seem
that Waltheof himself did not go to Hastings, and he,
together with Edwin and Morcar, received certain marks
of favour from the victorious William. The two former
were, however, soon in revolt, and in 1068 William visited
York, Lincoln, Huntingdon, and Cambridge.

In 1069 Waltheof aided a Danish attack on the east
coast, especially distinguishing himself at York, whereupon William pillaged the district, slaughtering, burning,
and laying waste the whole land. Waltheof, however,
submitted and was restored to his earldom, and shortly
after married Judith, the Conqueror's niece. In 1072 he
was made Earl of Northumbria; but in 1076, being
accused of conspiring against the king, he was arrested
and beheaded.

After the death of Waltheof the possession of the
earldom and castle of Huntingdon became a cause of
dispute between his heirs—the Scottish royal family and
the family of St Liz. First one party and then the other
gained the ascendency, until at last King Henry II levelled
the castle to the ground in 1175, thus putting a stop to
the constant fighting. The county being now at peace,
there is little to record until the reign of Henry VIII.

We may however mention that the town of Huntingdon, like most other towns, had at one time a strong

Jewish colony with a synagogue, but in 1289, when
Edward I expelled the Jews from England, the synagogue
was burnt down and their books sold. The valuable
Hebrew books being confiscated, many of them were
bought by the monks at Ramsey, who throughout the
middle ages were celebrated for their study of Hebrew
literature.

Huntingdonshire suffered like the rest of the country
from the three great plagues of 1349, 1361–2 and 1369,
so much so that the town of Huntingdon lost one quarter
of its inhabitants and is said never to have recovered its
former prosperity.

At the great upheaval known as the Reformation the
lands of the great Abbey of Ramsey and of the Priories
of St Neots, Huntingdon, and Hinchingbrooke passed to
Sir Richard Williams, alias Cromwell, who thus became
possessed of vast estates in this county.

At the time of the attempted invasion of England by
the Spanish Armada, in 1588, each hundred of the county
sent its contingent of 100 infantry, who marched to the
headquarters at Tilbury to be ready for any emergency,
one of the captains being Oliver, afterward Sir Oliver,
Cromwell. In addition, the county supplied 104 cavalry,
of which number Sir Henry Cromwell furnished 30.

The Cromwell family continued to hold a leading
position in the county and was in high favour at court.
It was Sir Oliver who entertained King James I at
Hinchingbrook, when on his way from Scotland to take
possession of the throne of England.

Robert Cromwell, a younger brother of this Sir Oliver

Hinchingbrooke

Cromwell, was the father of the famous Oliver Cromwell, who, with his cousin John Hampden, was one of the principal opponents in Parliament of the tax called "Ship-money."

A little later, when the colleges of Cambridge University made a grant of their plate to the King, Cromwell lay in wait with a band of horsemen at Lolworth Thicket, just outside the county, to intercept it on its way, but, under the guidance of Barnabas Oley, vicar of Great Gransden, the first portion of the plate was conveyed through the byways of Huntingdonshire and safely delivered to the King at Nottingham.

This brings us to the Civil War, when Huntingdonshire was attached to the Eastern Counties' Association, and Oliver Cromwell with his friends and neighbours rapidly obtained a leading position in affairs. How many of the foremost men came from Huntingdonshire and the adjoining counties may be gathered when we say that Cromwell, Edward Earl of Manchester, Edward Montagu (afterwards Earl of Sandwich), Valentine Wauton (one of the regicides), Henry Lawrence (the president of the council), and Stephen Marshall (a Parliamentary divine) were all Huntingdonshire men, and Colonel Desborough came from Eltisley, only just outside the county. The war did not greatly extend into Huntingdonshire, but on August 24th, 1645, King Charles, with a body of 2400 horse, after a skirmish at Stilton and another on the outskirts of Huntingdon, entered and occupied that town. Again on July 9th, 1648, a sharp fight took place at St Neots, when the Duke of Buckingham and the Earl of

Holland were defeated. Buckingham escaped; a small body of his men was pursued by the rebels as far as Houghton on the river Ouse, and it is said that they occupied the church of Little Paxton to block the road. The Earl of Holland was captured in St Neots and afterwards executed in flagrant disregard of the terms of surrender.

Cottage at Coppingford, where Charles I slept

Another interesting episode of the Civil War is the flight of King Charles to join the Scotch army at Newark, travelling by way of Ely, Earith, and Stukeley. He went to the house of Mr John Ferrar of Little Gidding, who conducted him to Coppingford, a neighbouring village, where he stayed the night of May 2nd, 1646. He spent Sunday there, going on in the evening to Stamford, and finally reaching the Scotch army at Newark on Tuesday, May 5th.

If some of the men of Huntingdonshire were prominent in the overthrow of King Charles I, it was reserved for one of them to be the prime mover in the restoration of Charles II, his son. Edward Montagu, who was an Admiral of the Fleet during the time that Oliver Cromwell usurped the control of affairs, joined with General Monk in inviting the King to return, and went over to Holland to convey him to England and was soon after made Earl of Sandwich.

Amongst prominent Royalists also during the Civil War was Sir Oliver Cromwell, the Protector's uncle, whose sons fought on the side of the King and, up to as late as 1687, flags captured from the Parliamentary forces are said to have been displayed in Ramsey church. Sir Oliver suffered severely for his loyalty, for it was during this period that the manors of Warboys and Upwood were sold by him to meet the heavy fines inflicted by the party in power. It will thus be seen that not only was the county divided in its allegiance, but members of the same family fought on opposite sides.

In the year 1745, when Charles Edward—the young Pretender—made his great effort to obtain the crown of England, Huntingdonshire showed its loyalty by contributing over £2000 and by raising a force to oppose the invader.

No further historical events especially connected with this county are recorded until the time of the Napoleonic wars at the end of the eighteenth century, when the

Government, thinking it advisable to keep the prisoners captured in these contests at a distance from the coast, erected barracks at Norman Cross, a mile north of Stilton, for their reception. A site of 40 acres was purchased in 1796, and building operations were pushed on so rapidly that, in March, 1797, all was ready for the accommodation of 8000 men, that is 6000 prisoners with their guards. The prison was practically an enclosed camp, consisting of four equal squares, formed by the intersection of two wide streets at right angles to each other. These four squares were again subdivided into four equal divisions, that is 16 divisions in all, and each surrounded by a lofty wooden palisade. At the end of each of the divisions stood a two-storey wooden building, roofed with red tiles, built to accommodate 500 men each, and in the centre of the prison was a high block-house. Thirty wells, each 100 feet deep, were sunk; the whole area of the central prison was 15 acres and was enclosed by a brick wall. The principal entrance faced the Great North Road—a gateway with pillars surmounted by a stone ball, now placed at the entrance to Yaxley churchyard.

In March, 1797, orders were issued that the French prisoners at Portchester should be removed to Norman Cross. On the 10th of April the first batch came via the "Dog and Doublet," a small port between Peterborough, Whittlesey, and Thorney. Orders were given that prisoners should only be landed at Yarmouth or King's Lynn. Those from King's Lynn were taken by barges and lighters by way of the Forty-foot, the Hundred-foot, the Pauper's Cut, and the river Nene to

Peterborough and marched thence to Norman Cross; others came through Cambridge and Huntingdon. The prisoners employed themselves while at Norman Cross in making models and other articles, using for the purpose their beef-bones, straw from their beds, and even their

Model of Ship, made from beef-bones by the French
Prisoners at Norman Cross

own hair, while others made lace. Many of these articles, especially the models of ships, show unsurpassable fineness of workmanship. They were sold at a special market held in the barracks every day. The Peterborough Museum contains many specimens of their work.

The soldiers of the garrison who died at Norman Cross were at first buried in Yaxley churchyard, but later in the special cemetery attached to the prison. After the Peace of Amiens, March, 1802, the prisoners left for Dunkirk at the rate of 500 a week. The depôt was empty and put up to let in the summer of 1802, but war again breaking out, it was called into requisition once more in October, 1803; from which time fresh batches continued to arrive till, at the end of 1808, there were nearly 6000 prisoners in the barracks. At the Peace of 1814, all prisoners were again liberated and the buildings used for a short time as a depôt for a small park of artillery and two regiments of cavalry.

In 1818 the place was dismantled; the sale commenced on the 2nd of October and continued for nine days, realising £10,000.

14. Antiquities—Prehistoric. Roman. Saxon.

We have no written records of Man as he first lived in our land long ages ago. Writing was an unknown art, and records—even if they had existed—could not have survived to come down to us. We therefore speak of this period as the Prehistoric—the time when the people of the past were unable themselves to record their story. Yet, though these sources of information are closed to us, we are able from the relics they have left behind them— the implements and weapons that they used, the bones of

the animals they fed upon, the structures they erected—
to form a fairly clear idea of these early peoples.

But this Prehistoric period, vast in its extent, has
for convenience sake been further subdivided. At first
the metals were unknown, or at least unused, and this
period is spoken of as the Stone Age, for it was
of flints and other stones that weapons and domestic
implements were mainly fashioned. Later, man learnt
how to get the easily-worked ores of tin and copper from
the rocks and by their admixture to form bronze. From
this, beautiful weapons and other articles were made, and
from the time of the discovery we date what is known as
the Bronze Age. Doubtless the ores of iron had long
been known, but how to smelt them was another matter.
At length the method was discovered, and mankind was
in possession of hard metal implements having great ad-
vantages for all purposes over those previously employed.
Thus the Iron Age began, and the early inhabitants of
Britain had arrived at this stage of civilisation when the
Romans came to our land.

We may now turn to a consideration of these various
epochs in their order. Firstly the Stone Age. This,
though a convenient term as covering all the period
before the advent of the Metal Ages, is too indefinite
both as to time and race, and hence it is usual to speak
of the Palaeolithic or Old Stone Age, and the Neolithic
or New Stone Age. The people of these two Ages were
very distinct, and most authorities hold that—at all events
in our land—a vast gap of time separated them, though no
such gap occurred between the later Ages. Palaeolithic

man, from various causes, ceased to inhabit what we now call Britain, and when the country was re-peopled it was by Neolithic man. Palaeolithic man lived in the days when the mammoth, reindeer, and hyaena roamed over our country; made leaf-shaped roughly-flaked flint weapons which were never ground or polished; cultivated no plants and tamed no animals; and built no monuments, graves, or houses. Neolithic man, on the other hand, learnt how to grind and polish his implements; was both a farmer and a breeder of stock; had many industries; and built megalithic monuments, houses, and graves—the remains of which survive to the present day.

Of these early people very few remains have been discovered in Huntingdonshire. A stone celt was found at Ramsey about 25 years ago, another stone implement was found in the gravel at Orton Longueville by Lady Huntley, and bronze battle-axes are from time to time disinterred in the fens.

It is supposed that there was a British settlement at Chesterton and another at Horsey Bridge, where log canoes, with iron and brass spear-heads and fishing implements, have been found. Huntingdon and God-manchester are almost certain to have been British sites, and, at the latter place, flint implements have been discovered. In 1757, in a tumulus near Chatteris ferry in the parish of Somersham, several skeletons were found together with an iron sword, the umbo of a shield, an urn and a glass vase. But there is on the whole curiously little of these periods.

Of Roman remains a very different tale may be told.

Palaeolithic Flint Implement
(*From Kent's Cavern, Torquay*)

Neolithic Celt of Greenstone
(*From Bridlington, Yorks.*)

The most important Roman station was Durobrivae between Chesterton and Water Newton on the Nene. Here the Roman road passes through a great square rampart, whose north side seems to have been fortified with a wall and the other sides only with banks of earth, but in digging a trench on the south side foundations of hewn stone and thick iron bars 10 feet long, as of a portcullis, were found.

When the turnpike road from Kate's Cabin to Wansford was made, it cut through the Roman cemetery, and many urns of different clays and forms were discovered, together with coins and several stone coffins with parallel sides, and one leaden coffin which weighed 400 lbs. Near Stibbington another burial-ground was found, not far from the river.

In 1754, near Chesterton, other coffins were found, one of them containing a skeleton, three glass lacrymatories, a small black seal, a coin of Faustina, one of Gordian, and some scraps of white wood inscribed with Roman and Greek letters. Near this place, but chiefly on the Northamptonshire side of the river, a great quantity of Roman pottery was made, and to it the name of Durobrivian ware has been given.

Godmanchester, on the Ouse, was an important Roman station at which three great roads met and passed over the river. It is commonly supposed to be the site of the Roman city Durolipons. Here numerous coins have been found, amongst them those of Augustus, Tiberius, Claudius, Nero, Vespasian, Trajan, Hadrian, Antonine, etc., but so far no pavements or buildings have been recorded.

The neighbourhood of Colne, Somersham, and Earith abounds in Roman remains. In 1731 a small urn full of Roman coins was turned up near the road leading from Somersham to Chatteris, and, upon subsequently digging near the spot, another urn, containing about 60 coins, was also found; and again, in 1824, another urn containing a large number of copper coins of Constantius and Constantine. In forming a cutting of the St Ives and

Roman Vase, found near Worlick

Somersham Railway a small bronze two-handled cup was discovered.

Another urn was unearthed at Colne, when digging pits to bury the cattle that died of the cattle plague, and various coins have also been found here.

Close to the village of Earith, in the year 1814, a very fine bronze statuette of Jupiter Martialis was brought to light, which is now preserved in the British Museum.

Prehistoric Dug-out Boat
(*Discovered in Warboys Fen*)

About 1824, a small bronze figure of Mercury was found in a field close to Hail Weston, and on an adjacent farm traces of a Roman camp are thought to have been discovered. Not far away is the Roman site now known as the Cony-geer, in the parish of Eynesbury; here numerous coins have occurred, especially of Domitian, Adrian, Carausius, and Constantine, and much pottery has been found here and in the neighbourhood. Other Roman remains have been discovered at Broughton, Sawtry, and Ramsey.

Of the period between the departure of the Romans and the later Saxon times we have very few remains.

At Woodston are two Saxon cemeteries, in which have been found skeletons and cinerary urns side by side, the skeletons not lying east and west and therefore pointing to a date anterior to the conversion of this part of England to Christianity, probably the sixth century. Amongst the objects found were spurs, spear-heads, knives, girdle studs, fibulae, glass beads and the usual East Anglian pottery. Probably to this same period may be assigned the burial-place at Elton, which contained very similar objects.

In a field at Great Paxton, about 150 yards west of the church, a great number of skeletons have been dug up; they were first discovered in 1820, and again when the railway was made in 1849–50.

Finally it should be recorded that in 1910 there was disinterred in Warboys Fen the remarkable prehistoric boat figured on the opposite page.

Silver Thurible of thirteenth century, found in draining Whittlesey Mere

15.   Architecture.   (*a*)   Ecclesiastical— Churches.   Monastic Houses.

The ancient architecture of England is generally divided into two main divisions, viz. Romanesque and Gothic. Of these, the Romanesque may be again divided into Saxon and Norman; and the Gothic into Early English, Decorated, and Perpendicular. The Saxons in England were not great builders in stone, but a few examples of their work have come down to us, and may be known by their rude and coarse masonry; by the very small windows with semicircular heads (or sometimes triangular or straight sided arches), and wide splays outside as well as inside; by flat pilaster strips used as ornament on the face of the walls; and by "long and short work," or an alternation of long vertical stones and flat slabs, in the quoins or corners of the walls or towers. No buttresses were used, and the work is that of craftsmen with an imperfect knowledge of stone construction.

The Norman Romanesque, which may be said to date from the Conquest, was, like its predecessor, a round arched style, very rude at first, but afterwards becoming more refined and more highly ornamented. Massive square towers were common. The walls were of great thickness; the vaulting semicircular; the windows were semicircular-headed, with a wide splay inside and hardly any splay outside; the pilaster strips and the "long and short work" disappear; and circular columns begin to be much more frequently used. As time went on the Norman masons became very expert in carving somewhat bold ornament, frequently very grotesque heads of birds

Saxon Crosses, Elton

and dragons, and other ornaments such as the billet, star ornament, and chevron, the latter being a sign of late date. The Norman style lasted for about 120 years, and

then towards the end of the reign of Henry II it began
to give way to a lighter and more elegant style, and there
was perfected the science of vaulting, by which the
weight is brought upon piers and buttresses.   This
method of building—the Gothic—originated from the
endeavour to cover the widest and loftiest areas with
the greatest economy of stone.   But the change was very
gradual, and the buildings which show a mixture of the
features of both styles are spoken of as Transitional.
These buildings generally have semicircular and pointed
arches used side by side, many of the more refined features
of the Norman style, and a profusion of carving, often
consisting largely of trefoil leaves.

Of the Gothic styles, the first, or Early English
as it is called, is perhaps the most beautiful of all.
The pointed arch had now reached perfection and the
semicircular arch was almost entirely discarded.   The
windows were at first long narrow lights with a pointed
head, hence called lancets ; the juxtaposition of two or
more of these side by side soon gave rise to tracery in
the heads, chiefly consisting of circles, but sometimes of
trefoils.   The ornament consisted of rather stiff but grace-
ful conventional foliage, chiefly of trefoil leaves on long
slender stems.   The mouldings were deeply undercut, and
frequently enriched with the dog-tooth ornament, which
may be taken as the distinctive feature of the style.

This style, which had become fairly established by the
year 1190, lasted until the middle of the reign of King
Edward I, when it gradually gave way to the second
Gothic style, called the Decorated.

It must be remembered that this name is more applicable to the larger and more costly works, and that the work of this period in the smaller parish churches is very simple and sometimes quite plain and homely, and is Decorated only in name. The distinctive features of this style are a greater richness and variety in its tracery, less conventional foliage, the leaves of the ivy and vine being very largely used; the mouldings not quite so deeply hollowed, and the abandonment of the dog-tooth ornament in favour of the ball-flower and the rosette. As time went on the tracery assumed a very flowing character, the carving became much richer and the mouldings flatter, and ogee mouldings became very prevalent both in the decorative arches and niches.

The Black Death stopped all building for some time. Then, with curious uniformity and quickness, the style termed Perpendicular—which is unknown abroad—developed after 1360 in all parts of England, and lasted with scarcely any change until 1520 or later. This style derives its name from its strongly-marked vertical and horizontal lines; the mullions of the windows run right up to the arch, and the circular and flowing lines of the earlier tracery disappear; the transom or horizontal bar is a frequent feature. The arches are flatter, and often four-centred. The mouldings became more and more shallow, and the ornament consisted largely of flat square paterae, flowing stems bearing vine leaves and grapes, and quaint grotesques. Towards the end of the style the Tudor ornaments, the rose and portcullis, were very largely used.

This period was one of great activity in church building; the men of the third estate, the sturdy yeoman, the prosperous merchant were fast accumulating wealth and they delighted to spend it upon re-building and ornamenting the House of God, and few indeed are the churches, not only in Huntingdonshire but throughout the whole of England, that do not present some mark of this great epoch.

It must be borne in mind that in the small country parishes, far removed from the wealth and influence of great monasteries or large towns, the masons and carpenters were often somewhat behind the times, and the dates of the parish churches may very reasonably be put some 20 years or more later than those named above for the various styles.

The architectural features in any particular locality are influenced by three main factors, viz. (i) the climate; (ii) the occupation and the wealth of the inhabitants; (iii) the materials available for building. And when we begin to apply the consideration of these factors as they influenced the county of Huntingdon we find (i) that the climate is temperate; there is need on the one hand for protection from the constant changes of weather—the wind and the rain of spring and autumn, the cold and snow of winter—which demand a roof over our heads and enclosing walls with windows and doors; while on the other hand the sun is never so powerful as to compel us to resort to broad verandahs, wide open loggias, or the roofless buildings of warmer climes. (ii) The inhabitants of this county have always been largely dependent upon agricultural pursuits for their living, those in the uplands being chiefly farmers growing corn and hay and rearing cattle and sheep, while in the lowlands or fens a considerable portion of the population adopted fishing as a livelihood. The inhabitants were therefore a hardy, prosperous folk, earning their bread by the sweat of their brow, and although well-off they were seldom rich and almost always of homely, simple habits of life. (iii) The county is almost entirely devoid of building stones, but being well traversed by rivers and water-ways, which were probably much more frequently used for the carriage of heavy materials than were the roads, it was no very serious matter to obtain the stone required for building their churches from the neighbouring counties of Northampton and Lincoln. At the same time the county was well wooded ; the oak tree, which gives us the finest of all the timbers used in building, being very plentiful.

We therefore find that the churches in this county are seldom very large, or very richly ornamented, as those of Norfolk and Suffolk ; nor are they so small as many in Sussex ; but are of medium size, simple and dignified, of pleasing proportion, and often full of beautiful woodwork as in the screen here given, which is in Wistow Church ; but with carving and decorative ornament very sparingly used.

Along the banks of the rivers and water-courses, especially in the western parts of the county approaching Northamptonshire, the churches often possess fine spires of wrought stone, which present very striking examples of the influence of materials upon buildings, for it is precisely to these churches that it would be most easy to convey the excellent building stone from Weldon in Northamptonshire, and from Ketton on the borders of Rutlandshire.

Oak Screen, St John Baptist, Wistow

There are no cathedrals in Huntingdonshire, but in olden days there were two abbeys, namely, Ramsey and Sawtry; four priories, St Mary Huntingdon, St Neots, St Ives, and Stoneley; one nunnery, Hinching-brooke; and a few smaller monastic establishments.

Of these a few portions of Ramsey Abbey remain, built up in the more modern residence of Lord de Ramsey, they are chiefly of Early English date ; Sawtry Abbey and the four priories have entirely disappeared, although their sites are known ; and the same fate has befallen the smaller houses with the exception of five arches of St John's

Ramsey Abbey

Hospital, now the grammar school, Huntingdon. The nunnery of Hinchingbrooke, altered and largely rebuilt, is now the seat of the Earl of Sandwich. And so it has come to pass that the mediaeval ecclesiastical architecture of Huntingdonshire is represented almost entirely by the parish churches.

Of the 96 ancient churches which still remain, and

the 22 others which are known at one time to have existed, 54 are mentioned in Domesday Book—an unusually

Arches in Refectory—Ramsey Abbey

high proportion—and it is therefore not surprising that many of the churches should exhibit ancient features. Nevertheless, Saxon workmanship only remains in two

cases, viz. in the west window (now built into the
modern tower) of Woodston Church, which is of very
rude construction, and widely splayed externally as well
as internally; and the columns and arches, and the
clerestory over them, at Great Paxton. But this latter
example, although clearly the work of Saxon builders, is
certainly post-Conquest in date; it exhibits the " long and
short " work of the Saxons in a rather curious manner, and
the windows are splayed outside and in, but at the same
time many features clearly point to a late period for its
erection.

It is evident that most of the churches mentioned in
Domesday Book were then simple little buildings of
timber, probably made of oak trees sawn down the middle
and then fixed vertically side by side, as may be seen to
this day at the interesting little church at Greenstead, in
Essex ; they were probably covered with a thatched roof,
and had no other floor than mother earth. In size they
no doubt consisted of a small nave and a still smaller
sanctuary; and in some cases a tiny porch, generally, if
not always, on the south side.

But when the Normans found themselves firmly
established in England they speedily set themselves to
rebuild these little churches in stone, and we therefore
find a very considerable amount of Norman work in the
churches, the best examples being the chancel at Ramsey,
the chancel and nave arcade at Fletton, the nave arcade
at Upwood, the tower at Farcet, doorways at Bury,
Stibbington, Little Paxton, Southoe, Covington, and Stow
Longa, and portions of numerous other churches. The

date of all this Norman work may be taken as between
1100 and 1170 A.D.

Of examples of the Transitional period the county
possesses the nave arcades at Barham, Bury, and Ramsey,
and a few doorways and fonts in other churches.

Of the Early English style Huntingdonshire has a
very large number of examples, prominent amongst which

Church of the Holy Cross, Bury

may be named the chancels at St Mary, Huntingdon,
Bury, Holywell, Somersham, Alconbury, and Molesworth;
the nave arcades at Wyton, Great Stukeley, and Abbots
Ripton; the towers at Alwalton, Bury, Warboys, Alconbury, Buckworth, and Chesterton; and doors, windows,
and fonts in numerous other churches.

Of examples of the earlier work of the Decorated style may be named the chancels at Brampton, Broughton, Stanground, and Spaldwick, and the tower at Holywell : and of the later work Orton Longueville, Swineshead, the chancel at Thurning, the tower at Spaldwick, and the aisles at Stow Longa, Great Staughton, Alconbury, and other churches.

Church of the Blessed Virgin Mary, St Neots

The more important examples in this county of the Perpendicular style are the churches at Wistow, Little Stukeley, St Neots, Buckden, Huntingdon All Saints, St Ives, and Great Gransden; and the naves at Ellington, Godmanchester, and Yaxley.

The Nave, All Saints, St Ives

16. Architecture. (*b*) Military—Castles.

Long before the coming of William the Conqueror there were castles in England, but by far the larger number of our English castles were built in Norman times, and the majority of these a very short time after the Conquest. Establishing themselves newly in the country it was incumbent upon the invaders to render themselves secure, and accordingly they set to work to construct efficient strongholds, with the result that many of them have survived to the present time. It is said that over eleven hundred castles were built in England about this period.

The design of these castles was more or less the same. Lofty and very thick walls with towers and bastions enclosed a considerable space of ground, and were surrounded whenever possible by a moat. This was crossed by a drawbridge at the entrance gateway which gave admission to the outer bailey or courtyard, where were the stables and offices. From this, another towered gateway led to the inner bailey which comprised the barracks, chapel, and other buildings, and—last resort when the rest had been carried by the enemy—the keep. This was a massive stronghold, generally square, and always provided with a well, in order that there might be no lack of water in the event of a siege.

Huntingdonshire was not a county of great castles and military works ; there was in early days a castle at Huntingdon, and the mound upon which it stood, overlooking the river and commanding the bridge (probably a ford in those days) still exists, but the castle itself was dismantled by King Henry II to put a stop to the frequent disputes for its possession between the rival claimants to the Earldom of Huntingdon.

At Kimbolton there was a castle or forcelet, which Leland describes as " doubly diked about and metely

Kimbolton Castle

strong." This castle belonged to the Mandevilles and their successors, the Bohuns and Staffords ; afterwards to the Wingfields who sold it to the Montagus, Earls and Dukes of Manchester. The first Duke of Manchester pulled down the old castle and built a large house on its site from designs by the celebrated playwright and architect Sir John Vanbrugh.

At Conington there was an ancient castle, once the
home of Turkill, the Danish Earl of Huntingdon, afterwards of Waltheof, the Saxon Earl, from whom it passed
to his descendants, including Prince David, brother of
Malcolm and William, Kings of Scotland, and the Bruces
and Cottons. The present house known as Conington
Castle was built by Sir Robert Cotton on a different site
from that of the ancient castle of the Bruces, and has been
largely rebuilt since then. It still retains some of the arches
of Fotheringhay Castle brought hither by Sir Robert Cotton.

17. Architecture. (c) Domestic—Manor Houses. Cottages. Inns. Bridges.

Huntingdonshire possesses its fair share of the old
manor houses and domestic habitations of the men of
old, but their fame is not spread abroad as is the case with
many another shire: there is no Hatfield or Penshurst, no
Charlecote or Ann Hathaway's cottage ; but scattered up
and down amidst her uplands and her fens is many an old
house that could tell its tale of adventure and tragedy.

The older and larger of these houses are built of stone,
and this is especially the case in the western parts of the
county, which are nearer the stone districts ; but the later
houses are built of brick, while the smaller ones are often
of timber and plaster. Foremost amongst these domestic
buildings of Huntingdonshire must be placed the ancient
palace of the Bishops of Lincoln at Buckden. A great
red brick tower, an entrance gateway and some boundary

walls are all that now remain, but it was once the home
of powerful prelates, and the quiet retreat of godly men.

Buckden Palace

Bishop Grosseteste, the great Englishman who defied the
Pope, died here in 1252, but the parts that now remain
are the work of Bishops Rotherham (1472–1480) and

Elton Hall

Russell (1480–1494). Bishop Williams spent considerable sums upon the house and gardens, but his work was
ruthlessly destroyed during the rebellion, and Bishop
Saunderson (1660–1663) had to expend a large sum to
make the building habitable again.

Hinchingbrooke, as we have seen, dates back to very
early times and was a nunnery. Elton Hall, the seat of

Old Gatehouse, Leighton Bromswold

the Earls of Carysfort, was probably built by the Sapcotes
in the fifteenth century. The fine gateway and some
parts of the present house date from at least the time
of King Henry VII, but other parts are of later and
varying dates. The Sapcotes sold the property about
the year 1600 to Sir Peter Proby, the ancestor of the
present owner.

At Leighton Bromswold is an interesting house of
red brick and stone, having small towers at the four
corners and great stone arches on the two principal faces;
this was built about 1603 by Sir Gervase Clifton, first
Baron Clifton of Leighton Bromswold, as an entrance

Cromwell House, Huntingdon

gateway to a large mansion which he intended to build,
but death put an end to his project, and the gatehouse,
after lying derelict for 300 years, has now been converted
into a vicarage-house for the parish.

The house at Huntingdon in which Oliver Cromwell
was born has been rebuilt more than once, and the present

house is entirely modern. Slepe was the ancient name for St Ives, and Slepe Hall where Cromwell lived in this town has been razed to the ground, but the front door has been preserved and re-erected at Bluntisham Rectory. A red brick barn at St Ives is still known as Cromwell's Barn.

At Upwood is a manor house of great antiquity; the actual building dates from about the seventeenth century, but there is no reason to doubt that it occupies the site of the house in which the renowned Duke Aylwyn was living when he founded Ramsey Abbey in 969, and which he gave to the monks at his death.

Of no less interest is the ancient house at Bodsey, near Ramsey, said to have been a hunting box of King Canute. The present house, although not so old as that, is certainly of great age.

Stukeley Hall, once the seat of the ancient family of de Stukeley, and afterwards of the Torkingtons; Gaynes Hall, formerly the home of the Engaynes; Washingley House, Orton Hall, Abbots Ripton Hall, Waresley Hall, and Great Gransden Hall occupy the sites of ancient houses, but have been modernised or rebuilt, and contain little remains of early date.

At Brampton is a large mansion, a seat of the Dukes of Manchester, inherited by them from the Sparrows and Bernards; it has recently been almost destroyed by fire, and the present house is entirely modern. In the same village is a small brick and plaster farmhouse, once the property and abode of the celebrated Samuel Pepys.

The house of the saintly family of Ferrars, at Little

Gidding, has gone from off the face of the earth ; but the quaint old plastered cottage at Coppingford, in which King Charles I passed the night when he was fleeing to join the Scotch army, still stands.

In the town of Huntingdon there are two houses connected with the poet Cowper, but they present no particular architectural features.

Pepys' House, Brampton

There are numerous other interesting old houses in various parts of the county, notably an ancient house at Woodhurst, and seventeenth century houses at Stibbington, Woodston, Hamerton, Warboys, and Toseland ; the last (p. 65) a very good specimen of the smaller red brick manor houses of the period.

There are several very interesting inns and coaching-houses, of which perhaps the best is the "Bell" at Stilton, a fine stone house bearing the date 1642 upon an added gable, and having a very large and elaborate wrought-iron sign. Of scarcely less interest is the "George" at Huntingdon, largely rebuilt but still retaining a quaint galleried courtyard, and some fragments of stone walling supposed to be the remains of a church

Huntingdon Bridge

dedicated to St George. The "Lion" at Buckden has a beautifully carved boss, representing the *Agnus Dei*, on an oak beam in the ceiling. Of others, the "Cross Keys" at St Neots is a rather early red-brick house ; while of later date are the "George and Dragon" at Buckden, the "Angel" at Stilton, the "Golden Lion" at St Ives, and the "Fountain" at Huntingdon. The

"Falcon" at the latter town was a place of some note in the seventeenth century.

The "Wheatsheaf," on Alconbury Hill, and the "Crown and Woolpack," at Conington, have both been converted into private houses; and the well-known "Haycock" at the foot of Wansford Bridge was until recently Lord Chesham's hunting box.

Nuns' Bridge, Huntingdon

Huntingdonshire is rather rich in ancient bridges. That at Huntingdon is an exceptionally fine specimen, probably erected immediately after 1294, in which year the earlier bridge was destroyed by a flood. At St Ives the bridge has the remains of an ancient chapel on its middle pier; the upper part has, however, been rebuilt with brick. The bridge at St Neots was built about the year 1589, to replace a more ancient one of timber.

One half of Wansford Bridge is in this county, but this part of the bridge was rebuilt in 1796. Other ancient bridges are the Nuns Bridge at Hinchingbrooke, Alconbury Bridge, Spaldwick Bridge, and a quaint timber foot-bridge at Hamerton.

18. Communications—Past and Present. Roads. Railways. Canals.

While we have very little authentic and definite information as to the mode of travel of our British forefathers, we do know that when the Romans came they found the British people possessed of horses and chariots, which they used in war. How far they made use of these for travel we cannot tell, but the Briton, on foot or on horseback, must have made his way from place to place, and all authorities agree that there was a road used by the Britons from Cambridge to Godmanchester, following very much the course of the present road connecting these towns. There is also a road called the Bullock Road, running from Huntingdon to Elton, and continuing northwards, which is believed to have existed in the days of the early Britons.

When the Romans came to a country one of the first works they undertook was that of road-making. This they did with a definite purpose. Having first occupied certain strategic positions, these positions were connected by roads, on which, at intervals, were stations of more or less importance, each occupied by a garrison. The roads

The Bell, Stilton (Old North Road, formerly Ermine Street)

were very well made—built, we might almost say—and were generally quite straight; a very good instance of which is furnished by the Roman road known as Ermine Street. From the spire of Godmanchester church, looking south, this road may be seen in a perfectly straight line, rising over hill after hill, but seeking always the shortest way to its objective. Ermine Street leads from London to Lincoln, entering Huntingdonshire from the south-east. Passing through Godmanchester, Huntingdon, Great and Little Stukeley, and Stilton, and between Alwalton and Chesterton, where was an important Roman station, it left the county near the village of Castor and ran thence straight to Stamford, this latter part being no longer in existence. Another Roman road is that from Cambridge to Godmanchester, following the old British track. It is part of the Via Devana, a road which connected Colchester with Chester. This road enters the county near Fenstanton, and joins the Ermine Street at Godmanchester, leaving it again near Alconbury, passing through the county in a north-westerly direction and entering Northamptonshire near Old Weston.

As time went on and stage coaches for public use began to take the place of the horse and "long wagon" used by our forefathers of the middle ages, many of these well-made and generally very straight Roman roads still continued to be used, while other roads were made to connect important towns. One very well-known example of the latter is the Great North Road from London to York, which traverses the county from end to end, entering it close to St Neots. It joins the Roman Ermine

The Old North Road (Ermine Street) near Chesterton

Street at Alconbury Hill, and leaves it again at Chesterton, entering Northamptonshire at Wansford.

Another good road runs from Huntingdon to Thrapston, and good cross roads connect the towns and villages of the county.

Carruther's *History of Huntingdon*, published in 1824, gives an account of what was considered the excellent

The Great North Road at Buckden

service of coaches at that date: "Few towns of the same size possess more convenient modes of conveyance to almost every part of the kingdom, as may be seen by the following enumeration. Daily coaches to London from the George Hotel, the Regent at 10 in the morning, Perseverance at 12, Boston Mail at 10 at night, Wellington at half past 10. From the Fountain, the Defiance at

10 in the morning, Edinburgh Mail at 11 at night, Defiance to Peterborough at 4 in the afternoon, Wellington to York at 11 at night, mails to Boston and Edinburgh at 4 in the morning. From the George, the Perseverance to Boston at 4 in the morning, the Rising Sun to Northampton at 8 and to Cambridge in the evening at 7. Mondays, Wednesdays and Fridays Blucher to Cambridge at 10 in the morning, Old Cambridge to Leicester at 9 in the morning, and to Cambridge at 6 in the evening; the Regent to Stamford at 4 in the afternoon. Van and waggons to and from London daily from Ashby's Waggon Office." The traveller by coach considered that he was moving at express speed if he covered twelve miles in an hour, while to-day we can reach London from Huntingdon in little over an hour.

The first railway constructed within the county of Huntingdon was that from Huntingdon to St Ives, which was opened on August 17th, 1847. This line was laid down by the East Anglian Railway Co., and was afterwards worked by the Eastern Counties Railway, whose system had been brought as far as St Ives from Cambridge and was opened at the same time. About the years 1850 to 1856 most of the trains on this branch were drawn by a horse, only one train in the day being worked by steam.

On March 1st, 1848, a branch of the Eastern Counties Railway was opened between St Ives and Wisbech, with a station at Somersham, but the lines from Somersham to Ramsey and from St Ives to Ely were opened later. All these lines are now worked by the Great Eastern, and the Great Northern and Great Eastern Joint Railways.

The main line of the Great Northern Railway Company which traverses the county from north to south, and now has stations at St Neots, Offord, Huntingdon, Abbots Ripton, Holme, and Yaxley, was opened on August 11th, 1850, but the first train to the north from King's Cross did not run until October 14th, 1852. The line from Holme to Ramsey was opened about 1865. On June 2nd, 1845, the London and Birmingham Railway opened a branch from Blisworth to Peterborough, which runs through Stibbington, Orton Longueville, and Woodston. The London and Birmingham, Grand Junction, and Manchester and Birmingham Railways were amalgamated in July, 1866, under the name of the London and North Western.

The line between Kettering and Huntingdon was made by the Kettering, Thrapston, and Huntingdon Railway Co., and opened on March 1st, 1866, being worked by the Midland Railway Co., who finally purchased the line in 1897.

Formerly great quantities of merchandise were carried by means of the water-ways with which Huntingdonshire was well provided. To this day the Fen district is intersected by a system of canals having landing stages at Holme Station, Yaxley, Higney, Great Raveley, Ramsey and other places, and connecting by means of Bevill's Leam with Whittlesey and Peterborough, and by way of Benwick with March, Wisbech, and Denver, and also by the Forty-foot river with the Old Bedford river at Welche's Dam; thus linking up the Fen system with the Ouse. The north-west corner of the county has water commu-

St Ives Bridge

nication, by means of the Nene, with Northamptonshire, through which, in the olden days, the celebrated Barnack Stone was brought to the large abbeys of the Fens; and a branch of the Nene passing through Whittlesey connected that river with the Fen water-ways.

The Ouse at Hemingford Grey

The southern part of the county is traversed by the River Ouse, which rises in Northamptonshire and passes through the counties of Northampton, Oxford, Buckingham, and Bedford. It enters Huntingdonshire near St Neots, flows through the towns of St Neots, Huntingdon, and St Ives, and leaves the county at Earith, where it joins the great system of the Bedford Level.

At one time the Ouse was navigable to Bedford, but the navigation does not now extend above St Ives.

19. Administration and Divisions— Ancient and Modern. Political Divisions.

We do not know much as to the administration of justice in England in early days. The Iceni, of whose territory Huntingdonshire formed part, lived under what was doubtless the somewhat arbitrary rule of their kings and chiefs; and the strong arm of the Roman conqueror kept order later. Then after the Romans left the island, the Angles soon took possession of this part of England and inaugurated their own system of government. Though the government and administration of justice must have been frequently carried out under difficulties owing to the unsettled state of the country through Danish invasions, it is quite clear that each tribe or district had its leader, who was elected to his office, and who later came to be spoken of as king.

In later Anglo-Saxon times we find the king exercising supreme authority, and England divided into counties; over each county the king appointed an earl or ealdorman, who for the administration of justice held a Shire Moot or Court twice every year, in May and October, at which all the thanes of the county had a right to be present and to vote. The executive officer appointed to carry out the decrees of the Court, and to take charge of prisoners, was called the Shire-reeve or Sheriff, and in the frequent absences of the Ealdorman he presided at the Shire Moot.

Besides these Shire Courts there were also Hundred Courts, presided over by the Sheriff, and all the freeholders of each hundred were required to attend the Sheriff in the courts of that hundred and to render him any assistance he required in the administration of justice. Each hundred probably consisted at first of one hundred free families, or perhaps comprised such an area as should furnish one hundred men capable of bearing arms, and in its turn each hundred was divided into townships, corresponding to our present parishes. Each township had its *gemot* or moot, where every freeman could appear. It made laws for the township—the *by* laws, or town laws—and was held whenever occasion arose.

If any suitor wished to appeal to a higher court, he had the right to appeal to the King, but the expense of such an appeal was so great as to be almost prohibitive.

Little change was made in the administration of justice after the Conquest till, in the reign of Henry II (A.D. 1176), an attempt was made to obviate this difficulty of appeal, and " Justices in Eyre," or Itinerant Justices, were appointed to make their circuits throughout the kingdom, and to hear and decide causes in the King's name. This system continues in but slightly different form to the present day.

Sometimes a district of considerable extent was cut off

from the rest of a county and called a "Liberty," within which some local magnate exercised judicial powers. The Liberty of Ramsey formed such a district in this county, comprising the parish of Ramsey and parishes and parts of parishes adjacent. Within this Liberty the abbot of Ramsey was responsible for maintaining order, and had power to inflict fines, imprisonment, and even the death penalty ; he had his own gaol ; his steward and others held courts, and the records of some of those held in the early part of the fourteenth century are still extant. Assault, murder, fishing in waters belonging to the abbot, deer-stealing, and other offences were investigated, and sentences passed on the offenders when found guilty.

In many cases, too, the lord of a manor had considerable power, licensing butchers, bakers and others to exercise their trades, and fining those who "broke the assize of ale," or "of bread," as the case might be. Records still preserved also show that tithings existed, that is, each ten freemen or householders were bound one for the other, so that each was held responsible for the good conduct of the rest—an excellent method of maintaining order.

At the present day the chief officer of the county is the Lord Lieutenant, and under him the High Sheriff.

For the selection of a High Sheriff, Huntingdonshire is joined to Cambridgeshire, the rule being to choose the Sheriff from Huntingdonshire one year, the next year from Cambridgeshire, and the next from the Isle of Ely.

Assizes for the county are held at Huntingdon, the

county town, for the trial of prisoners charged with the most serious offences, such cases being tried before a judge who travels on circuit, holding courts at various places in the King's name ; these are held twice a year.   To try less serious charges, Quarter Sessions are held at Huntingdon four times a year, while courts called Petty Sessions

The Town Hall, Huntingdon

are held in the different magisterial divisions, usually once a fortnight, presided over by Justices of the Peace who hold the King's commission for that purpose.

Since April, 1889, when the Local Government Act of 1888 came into force, the administrative business of the county has been in the hands of the County Council, which consists of a chairman, thirteen aldermen, and thirty-nine councillors ; and the elementary schools were placed under the control of the same authority in 1902.

An Act of 1894 established District Councils, subsidiary to the County Councils, to deal with parish roads, bridges and other matters, so that now the administration of the Poor Laws, the repair of the roads and bridges, and the management of the schools all come under the control of the latter body. A committee of the County Council and Justices of Quarter Sessions control the police and appoint the Clerk of the Peace.

There are in the county three boroughs, viz. Huntingdon, Godmanchester, and St Ives, which are each governed by a mayor and aldermen instead of by a District Council.

Huntingdonshire is divided into three Poor Law Unions—Huntingdon, St Ives, and St Neots. The St Ives Union includes some parishes in Cambridgeshire, and the St Neots Union some in Bedfordshire and Graveley in Cambridgeshire ; while some Huntingdonshire parishes are included in the Unions of Caxton, Cambridgeshire; of Oundle, Peterborough, and Thrapston in Northamptonshire; and of Stamford in Lincolnshire.

The county contains four hundreds, Norman Cross, Hurstingstone, Leightonstone, and Toseland ; three municipal boroughs, Huntingdon, Godmanchester, and St Ives; and 105 civil parishes. Ramsey, Kimbolton, and St Neots are market-towns ; and for several centuries Yaxley and Spaldwick also were, but the markets have now been discontinued.

Two members were returned to Parliament for the

borough of Huntingdon, and two for the county as early as 1296: this arrangement continued down to the year 1867, when the representation of the borough was reduced to one member. By the Act of 1885, representation was still further reduced, and the county now returns two members, the Borough of Huntingdon being merged in the southern division of the county.

20. Roll of Honour.

Let us now glance at some persons of note connected with Huntingdonshire by birth or residence.

King Canute resided at Bodsey and visits of several royal personages to Ramsey Abbey are recorded, among them those of Henry I and his queen, Edward II and his queen, and Edward III and his mother. Queen Philippa (Edward III's Consort) stayed at Ramsey Abbey in 1337. And at Kimbolton Castle, which was her dower-house, lived Queen Catharine of Aragon after her divorce ; there she died, and thence her body was taken for interment in Peterborough Cathedral.

James I stayed with Sir Henry Cromwell at Hinchingbrooke on his way from Scotland to take possession of his kingdom of England, and was much pleased with the sumptuous entertainment provided for him. Little Gidding, the home of the Ferrars, was visited more than once by Charles I, and at Coppingford Hill, it will be remembered, stands the cottage where he spent a night on his journey to seek the protection of the Scotch army (p. 72).

Amongst distinguished men to whom the county has given birth must first be mentioned Oliver Cromwell, who was born at Huntingdon in 1599. He was a scion of a distinguished Welsh family, named Williams, one

Oliver Cromwell

member of which, Morgan Williams, followed the fortunes of Henry VII and coming to Court, married a sister of Thomas Cromwell, Earl of Essex. His descendants adopted the name of Cromwell. Oliver Cromwell was educated at Huntingdon grammar school under Dr Thomas Beard, and at Sidney Sussex College, Cambridge. He was at one time a farmer at St Ives and

Dr Beard

later resided at Huntingdon. Elected M.P. for Huntingdon in 1627 he supported his cousin, John Hampden, in his resistance to the payment of ship-money. When the Civil War broke out, Oliver raised a troop of cavalry in the Parliamentarian interest, and soon showed conspicuous ability as a military commander, and from this point he rapidly advanced to the position of " Protector." During this part of his life his history is bound up with that of the nation.

Several supporters of Cromwell came from the county and the neighbourhood, Sir Henry Lawrence, president of the council, was a native of St Ives, Valentine Wauton, one of the regicides, was of Great Staughton, and Stephen Marshall, the celebrated Presbyterian divine, of God-manchester.

Among soldiers we may place Waltheof, the son of Siward and brother-in-law of Duncan, King of Scotland, who, though he does not seem to have been present at Hastings, distinguished himself at York, and, it is said, slew 100 Normans with his own hand. He subsequently surrendered to the Conqueror, whose niece he married, but he was eventually executed at Winchester on a charge of treason. Waltheof was an ancestor of our present King. John Tiptoft, Earl of Worcester, a strong supporter of the Yorkist party in the Wars of the Roses, was born at Everton in 1471. He was attainted, and, being found hiding in a forest near Huntingdon (probably Waybridge close to his estate at Woolley), was beheaded. Sir Edward Montagu, Earl of Manchester, ancestor of the Earls and Dukes of Manchester, is said to have been born at Kimbolton Castle, and, like Oliver Cromwell, was educated at Sidney Sussex College, Cambridge. During

Edward, 1st Earl of Sandwich

the Civil War he took an active part on the side of the
Parliament, but later was instrumental in promoting the
Restoration of Charles II.

As distinguished at sea as his cousin the Earl of
Manchester on land, Edward, Earl of Sandwich, was
associated with Monk in the Restoration, and was killed
in action against the Dutch in Southwold Bay in 1672.

Among historians, Henry of Huntingdon, no doubt a
native of that town, was Archdeacon of Huntingdon from
about 1109 till 1154. He wrote a History of England from
the time of the Romans to the accession of King Henry II
(1154), and *An Epistle to Warin, the Briton*, containing
an account of the ancient British kings from Brute to
Cadwallader. Besides the above he wrote other works,
and is worthy of note as being the first of our national
historians as distinguished from chroniclers. Roger de
St Ives, an Augustine friar, wrote against the Lollards
about 1390, and William de Ramsey, a native of Ramsey,
and Abbot of Peterborough in the reign of Edward IV,
was the author of a life of St Guthlac in verse, and other
works.

Besides the divines previously mentioned were Robert
Grosseteste, vicar of Abbotsley and afterwards Bishop of
Lincoln, an ardent Church reformer; Gilbert de Segrave,
a distinguished member of the Segrave family, rector
of Fenstanton, and made Bishop of London in 1313;
William de Whittlesey, rector of the same parish,
Archdeacon of Huntingdon, and in 1368 appointed
Archbishop of Canterbury. William de Sautré, a native
of Sawtry and a London clergyman, was the first man

burnt in England as a heretic under the Act passed in the reign of Henry IV. Mikepher Alphery, rector of Woolley 1618–65, was a member of the Russian royal family Alferiev and was himself offered the crown of Russia.

A distinguished man during the Commonwealth was Stephen Marshall, already mentioned as one of Cromwell's men. He was born at Godmanchester and educated at Emmanuel College, Cambridge; he became a Presbyterian, and was one of the chaplains of the House of Commons during the Civil War. Marshall was also one of the five authors of *An Answer to a Book entitled An Humble Remonstrance.* This treatise was published as written by "Smectymnuus," the word being composed of the initials of its authors' names, S-tephen M-arshall, E-dmund C-alamy, T-homas Y-oung, M-atthew N-ewcomen, and W-illiam S-purstow.

Nicholas Ferrar, the friend of George Herbert, held a unique position in the county and indeed in the whole country. The son of a London merchant, he took his degree at Cambridge, and afterwards became Deputy Governor of the Virginia Company. On retiring from public life he took up his abode at Little Gidding, where he founded a sort of private religious community, familiar to readers of *John Inglesant.* His mother, his sister, and her husband with their children were among the members, and spent their time continuously in various religious exercises. Mr Ferrar composed moral essays and short histories which were read aloud in rotation by the members, while the others were occupied in bookbinding

and kindred occupations. Nicholas Ferrar composed
several works for the use of Prince Charles, among
others a *Harmony of the New Testament*. The Parliamentary party plundered the church of Little Gidding

Nicholas Ferrar
(*From the Portrait in Magdalene College, Cambridge*)

and the mansion of the Ferrars, and thus perished most
of the works of this earthly saint. This act was com-
mitted just before the execution of Charles I, so that
Nicholas Ferrar, who died in 1637, did not live to see
the destruction of his work.

Huntingdonshire has furnished London with more than one Lord Mayor. Sir John Gedney, born at St Neots, held that office in 1427 and 1447 ; Sir Robert Drope, also of St Neots, in 1474; Sir Ambrose Nicholas, born at Needingworth, in 1576; and Sir Wolstan Dixie, a native of Great Catworth, in 1585.

Born February 23, 1632 (though it is not known whether at Brampton or elsewhere) and often a visitor to the county, few have contributed more to the amusement and the instruction of posterity than Pepys—the author of the *Diary* that, so long as the English language lasts, will be read. Beginning his journal from the Restoration he continues it for less than ten years, photographing Charles II's Court and the inner history of the period with marvellous detail, and revealing himself now as a hard-working and conscientious Secretary of the Admiralty, now as a mere voluptuary. Musician and playgoer, virtuoso, book-lover, traveller and scientist, with an insatiable interest in almost every conceivable subject, he covers a most important period which includes the Plague and the Fire of London, and his *Diary* is a mine of information for the historian.

Among the men of note we must not forget Sir Robert Cotton, the celebrated antiquary and founder of the Cottonian Library in the British Museum, who was born on the 22nd of January, 1570, at Denton. Cowper, the poet, though not born in the county, resided for two years at Huntingdon with the Unwins.

Bust of Sir Robert Cotton, from his tomb in
Conington Church

21. THE CHIEF TOWNS AND VILLAGES
OF HUNTINGDONSHIRE.

(The figures in brackets after each name give the population in
1901, and those at the end of the sections give the references
to the text.)

Abbotsley (329), five miles south-east from St Neots station.
Formerly belonged to the kings of Scotland, statues of two of
whom are on the church tower, together with those of two of the
kings of England.

Alconbury (543), four miles north-west from Huntingdon.
Here is the well-known "Alconbury Hill" on the great coaching
road to the north. A short distance south of the village is
Matcham's gibbet, where was executed Jarvis Matcham, whose
murder of Benjamin Jones a drummer-boy on the 19th August
1780, and his subsequent execution on the 2nd August 1786, is
the theme of Barham's poem "The Dead Drummer of Salisbury
Plain" in the *Ingoldsby Legends*. (pp. 95, 96, 108, 109, 111, 113.)

Alwalton (225), four and a half miles west-south-west of
Peterborough. The church has a fine Early English tower.
From quarries in the neighbourhood was obtained the marble
known by the name of the village, and largely used in Peterborough Cathedral and neighbouring churches. (pp. 31, 95.)

Bluntisham (965), four and a half miles north-east of St Ives. Within this parish is the hamlet of Earith, where was found a bronze statue of Jupiter Martialis now in the British Museum; there is also an ancient fort known as "the Bulwarks," possibly of Cromwellian date. (pp. 32, 49, 105.)

Brampton (1020), two miles south-west of Huntingdon. Samuel Pepys, the celebrated Diarist, had a house here which still

Cottages at Brampton

exists. The judge Sir Henry Hawkins took his title of Baron Brampton from this place. (pp. 34, 96, 105, 106.)

Broughton (264), five miles north-east from Huntingdon. The Abbot of Ramsey made this place the head of his Barony. The church has a curious fresco over the chancel arch. (pp. 83, 96.)

Buckden (1021), four miles south-west from Huntingdon. This village on the Great North Road possesses the remains of an ancient palace of the Bishops of Lincoln, and the church contains monuments to several of them who are buried here. (pp. 96, 100, 107, 113.)

Buckworth (164), eight miles north-west from Huntingdon. The church has a fine Early English tower and spire. (p. 95.)

Bury (370), one mile south from Ramsey. The church possesses a fine specimen of an oak lectern of the Decorated period, and also an ancient font. Formerly there was a chapel dedicated to "Our Lady of Bury."

Catworth, Great (444), three and a half miles north-west from Kimbolton. The birth-place of Sir Wolstan Dixie, Lord Mayor of London. (pp. 31, 131.)

Chesterton (118), five and a half miles west from Peterborough, derives its name from the Roman station, Durobrivae, which is in the adjoining parish. It was the seat of the ancient family of Devill; and afterwards of John Driden, cousin of the poet. (pp. 46, 78, 80, 95, 111, 112, 113.)

Colne (260), five and a half miles north-east from St Ives. The ancient church fell down in 1896, and a new one was erected in the middle of the village. (pp. 49, 81.)

Conington (268), two miles south-west from Holme station. Conington was the seat of the renowned Earl Waltheof, by whose daughter it passed to the royal house of Scotland, and so to the mother of Robert Bruce, the later King of Scotland.

From the Bruces the manor passed to the Cottons, of whom the well-known antiquary Sir Robert Cotton founded the library bearing his name in the British Museum. (pp. 100, 108.)

Coppingford (34), eight miles north-west from Huntingdon. There was once a church, the site of which is still known, but it has long since been demolished. (pp. 72, 106, 123.)

Denton (78), four miles west from Holme station. The manor was once the property of King Robert Bruce. Here was born the celebrated antiquary Sir Robert Cotton, on the 22nd of January 1570, his parents finding their house here more commodious than their residence at Conington.   (p. 2.)

Elton (674), eight miles west-south-west of Peterborough. Elton Hall was anciently the property of the Sapcotes, and later became the seat of the earls of Carysfort.   It contains a magnificent library of early printed books, and here are preserved a thurible and censer found at the draining of Whittlesey Mere, doubtless formerly the property of Ramsey Abbey.   (pp. 9, 23, 40, 83, 86, 102, 103, 109.)

Eynesbury (1091), adjoining St Neots, takes its name from Ainulf a Saxon hermit, in whose honour a monastery was founded here in the reign of King Edgar.   (p. 83.)

Farcet (1023), three miles south-east from Peterborough. A great part of the parish is fen land, and formed part of Whittlesey mere.   Here are large brickfields.   (pp. 55, 94.)

Fenstanton (886), two miles south from St Ives.   The chancel of the church is a very fine specimen of Decorated architecture.   (pp. 49, 111, 128.)

Fletton (1833), one mile south from Peterborough.   Fletton is now practically a suburb of Peterborough, although it has an Urban District Council of its own.   Here are extensive brickfields; Fletton bricks are known and used all round London and in the south midlands.   (pp. 30, 47, 52, 55, 94.)

Great Gidding (337).   Twelve miles north-west from Huntingdon.   Here are the kennels of the Fitzwilliam hunt. It is interesting to note that in the thirteenth century, John Engayne held Great Gidding by tenure of hunting the wolf, fox, cat, and other vermin.

Little Gidding (39).   One mile south-east from Great Gidding.   Little Gidding was the home of the saintly family of Ferrar, friends of the poet George Herbert, and the church in

St John's, Little Gidding

which they worshipped still contains many relics of them, for instance the brass lectern and font, the oak stalls, and some embroidery. (pp. 72, 106, 123, 129, 130.)

Glatton (189), ten and a half miles north-west from Huntingdon. At one time a rather important village, and the seat of the Castle and Sherrard families. (p. 61.)

Godmanchester (2017). Half a mile south from Huntingdon, which it adjoins. By some considered the site of the Roman station Durolipons. Godmanchester is a very ancient borough, its first charter dating from the time of King John; and at one time it joined with Huntingdon in returning two members to Parliament. When the king came into the borough the burgesses used to meet him with twelve ploughs and their teams of horses. At one time there was a considerable tanning industry carried on here. (pp. 30, 46, 47, 55, 57, 78, 80, 96, 109, 111, 122, 126, 129.)

Great Gransden (504). Seven miles south-south-east from St Neots. A considerable village, of which the celebrated Barnabas Oley, who was largely instrumental in conveying the Cambridge college plate to King Charles I, was vicar; he built five almshouses here, which still exist, dated 1679. The church is a fine specimen of the Perpendicular style of architecture; and contains a pulpit which is said to have come from Great St Mary's Church, Cambridge. (pp. 32, 71, 96, 105.)

Hail Weston (251). Eight miles south-west from Huntingdon, two miles north-west from St Neots. Here are two mineral springs, which were in high repute for medicinal purposes in the time of Queen Elizabeth, and probably long before. What is chiefly remarkable is that the two springs, although near together, are of quite different character. (p. 83.)

Hartford (410), one mile east from Huntingdon. The ancient royal forest of Sapley is now included in this parish. (p. 32.)

Hemingford Abbots (345); three miles east from Huntingdon; and **Hemingford Grey** (852), four miles east from Huntingdon. Two pretty villages on the Ouse. At the latter lived, in the reign of King George II, the two beautiful Miss Gunnings, one of whom married the Earl of Coventry, while the other became first the wife of the Duke of Hamilton and secondly of the Duke of Argyll. (pp. 2, 32, 49.)

Hemingford Mill

Hilton (275), four miles south from St Ives. On the village green a curious maze is cut in the turf. (p. 9.)

Holme (658). Eleven miles north from Huntingdon. An extensive parish, a great part of which, until 1850, was part of Whittlesey Mere. There is a station on the Great Northern Railway, and a branch line runs from here to Ramsey. (pp. 22, 54, 61, 62, 115.)

Holywell (653). Two and a half miles east from St Ives. The church is an interesting building of mixed styles; in the churchyard is an ancient well or spring, which gives name to the place. (pp. 10, 51, 95, 96.)

Houghton (324), three and a half miles east from Huntingdon. Here is an ancient and very picturesque watermill, on the Ouse. (pp. 55, 56.)

Huntingdon (4349). Fifty-nine miles from London on the Great Northern Railway; is the county town. Huntingdon is

New Grammar School, Huntingdon

situated on the banks of the Ouse, and has a very fine old bridge over the river. It is a borough by prescription and at one time returned two members to Parliament. In early days it had a castle, the site of which may still be seen. It is said that at one time there were twelve churches, but the town was decimated by the Black Death, in 1349, and in the seventeenth century only four of these churches remained. Now there are but two, although the four parishes are still kept distinct. (pp. 2, 12, 13, 17, 24, 47, 55, 68, 69, 75, 78, 92, 95, 98, 104, 106, 107, 108, 109, 111, 113, 114, 115, 118, 120, 121, 122, 123, 124, 125, 126, 131.)

Old Hurst and **Wood Hurst** (103 and 253), about four miles north from St Ives, are two small villages which appear to have given name to the hundred of Hurstingstone. A curious stone on the road-side near the former village was no doubt once used to mark the meeting place of the Hundred Court. (p. 122.)

Keyston (175), seven miles north-west from Kimbolton. The seat of Aluric the Saxon sheriff of Huntingdon; and, later, the property of the de Ferrers family. (p. 12.)

Kimbolton (915), ten and a quarter miles west from Huntingdon. Now a large village, but formerly a market-town. Here is the seat of the Duke of Manchester, formerly the castle of the Mandevilles, Bohuns, Staffords, and Wingfields. Queen Catharine of Aragon, after her divorce from Henry VIII, passed the end of her life here, and died in the castle. At Stoneley, a hamlet in the parish, was a small Augustinian priory. (pp. 9, 12, 47, 99, 122, 123, 126.)

Leighton Bromswold (287). Five miles north of Kimbolton. A pretty village finely situated upon a hill. George Herbert was prebendary of this place, and almost rebuilt the church. Sir Gervase Clifton, first Baron Clifton of Leighton Bromswold, began to build a large house within a moated enclosure; the gatehouse, a fine building of the early years of the seventeenth century, has now been converted into the vicarage house. (p. 104.)

Offord Cluny (207), three miles south by west of Huntingdon, on the Ouse, with a railway station and an Early English church. **Offord Darcy** (338), lies about half a mile to the south. The whole, probably, of both parishes belonged at one time to the Abbey of Cluny. Lying on the high road between St Neots and Huntingdon these are favourite fishing resorts. (pp. 55, 115.)

Orton Longueville (247). Two miles south-west from Peterborough. Here is the seat of the Marquis of Huntly. The house is modern, but an ancient gateway remains, under which it is said that the Duke of Somerset passed on the night before the execution of Mary Queen of Scots. (pp. 78, 96, 115.)

Orton Waterville (276), three miles from Peterborough. Sometimes called Cherry Orton, from the great quantity of cherries formerly grown there.

Great Paxton (244). Seven and a quarter miles south from Huntingdon. The church is an interesting example of Saxon workmanship executed after the Conquest. The village belonged to Earl Waltheof, and he probably commenced to build the church just before his death in 1076. (pp. 83, 94.)

Ramsey (4823 civil parish, 3465 ecclesiastical parish). A market-town, ten miles north-north-east from Huntingdon, and the centre of a thriving agricultural and vegetable-growing district. Here was the wealthy abbey of Ramsey, founded in 969, and now the seat of Lord de Ramsey. (pp. 2, 13, 21, 26, 30, 35, 37, 47, 55, 60, 66, 69, 73, 78, 83, 91, 92, 93, 94, 95, 105, 114, 115, 120, 122, 123, 128.)

Ramsey St Mary (1245), two miles north-west from Ramsey, is built upon the site of Ugg mere.

St Ives (2910). An ancient market-town which, in 1874, was incorporated as a borough. It has a railway station to which the Great Eastern, Great Northern, and Midland Railways run trains. There are extensive markets for horses and cattle. Formerly there was a priory here, subordinate to the great Abbey

The Waits, St Ives

of Ramsey. On the fine bridge are the remains of a bridge chapel. (pp. 10, 13, 25, 47, 55, 56, 81, 91, 96, 97, 105, 107, 108, 114, 118, 122, 126.)

St Neots (2798). A market-town, with a station on the Great Northern railway, nine miles south-south-west from Huntingdon. The market-place is a very fine open square, and there is a good market. The church, wholly of the Perpendicular period, is one of the finest in the county. St Neots Priory, originally founded by Earl Leofric, and refounded by Lady Roisia de Clare in 1113, was a cell to the Abbey of Bec, in Normandy. Hither were conveyed from Cornwall the relics of St Neot. There is a fine bridge over the Ouse. (pp. 24, 30, 55, 56, 64, 69, 71, 72, 91, 96, 107, 108, 11:, 115, 118, 122, 131.)

Sawtry (848). Eight miles north-north-west of Huntingdon. Sawtry consists of two parishes, All Saints and St Andrew, and the extra-parochial district of Sawtry Judith. In the last stood the Abbey of Sawtry, founded by Simon St Liz, a grandson of the Countess Judith; it belonged to the Cistercian Order. (pp. 83, 91, 92, 128.)

Somersham (1893). An old town with an Early English church five miles north-east from St Ives. In this parish once stood a palace of the Bishops of Ely. (pp. 32, 49, 78, 81, 95, 114.)

Southoe (213). Two miles north of St Neots. At the time of Domesday Southoe was the seat of Eustace de Lovetot, Sheriff of Huntingdonshire. The church has a very richly carved Norman doorway in the south wall. (pp. 61, 94.)

Spaldwick (249). Seven miles west from Huntingdon. At one time a market-town, but now only a village. It was given by the Abbot of Ely to the Bishop of Lincoln in recompense for his loss when the Abbacy of Ely was made into a Bishopric. The church has a very fine tower and spire, and a pretty early Decorated chancel. In the village is an ancient stone bridge. (pp. 57, 58, 96.)

Stanground (1461), one and a quarter miles east from Peterborough. A large village where once flourished an important fishing industry. (pp. 47, 96.)

Great Staughton (746). Nine miles south-west from Huntingdon. A large village in which were the seats of several important families. Staughton Manor was the property of the Criolls, Cretings, and Wauton families; Valentine Wauton was one of the regicides, and forfeited his estates. Gaynes Hall was the seat of the Engaynes; and Place House was occupied by Oliver Leder. Staughton House was the seat of the Baldwins, Onslows, and others. (pp. 96, 126.)

Stibbington (426). Nine miles west from Peterborough. Here is a fine seventeenth century house known as Stibbington Hall. The well-known posting inn, the "Haycock," at Wansford is also in this parish, and so is the southern half of Wansford bridge. (pp. 30, 80, 94, 106, 115.)

Stilton (535). Six miles south-south-west of Peterborough. At one time an important coaching place; there were two large coaching inns here, and the well-known Stilton cheese takes its name from this place, having been sold to the travellers at the inns. (pp. 74, 107, 110, 111.)

Great and Little Stukeley (346 and 200). Two villages, respectively two miles and three miles north-west of Huntingdon. The Norman church at Little Stukeley was built by Henry of Huntingdon, the historian. (pp. 72, 95, 96, 105, 111.)

Toseland (174). Three miles north-east of St Neots. A small village, but the centre of the hundred of Toseland, to which it gives its name. The name is derived from a Dane, named Toli, who was slain at Tempsford in 921, and is one of the very rare cases in which the person who gave his name to a village can actually be identified. There is a small Norman church and a red-brick manor house; and a large stone in the churchyard is said to be the ancient hundred-stone. (pp. 65, 66, 106, 122.)

Upwood (386). Two miles south-west of Ramsey. Here was the seat of Duke Aylwyn, founder of Ramsey Abbey (969), who gave his manor house to the Abbey at his death. The present manor house doubtless stands upon the same site, and was at one time the residence of a branch of the Cromwell family. (pp. 73, 94, 105.)

Warboys (1758). Seven miles north-east from Huntingdon. The church is chiefly of Early English date, and has a fine tower

St Mary Magdalene, Warboys

and spire at the west end; the chancel arch is Norman. (pp. 30, 73, 82, 83, 95, 106.)

Waresley (216), seven miles south-east from St Neots, was a seat of a branch of the Engayne family; afterwards of the Hewetts, Byngs and Duncombes. The old church was pulled down in 1856 and rebuilt on a new site. (pp. 9, 32, 105.)

Water Newton (105). Six miles west by south from Peterborough.   In this parish is the site of the Roman station, Durobrivae, on the Ermine Street: the great square site of the

St Peter's, Yaxley

camp may still be plainly traced.   Many Roman coins are found here, and not far away are two Roman cemeteries.   (pp. 26, 30, 31, 80.)

Wistow (352), seven miles north-north-east from Huntingdon. The church is a good specimen of Perpendicular architecture; it has the very unusual feature of a cusped rear-arch to the clerestory windows. There are good oak screens, and some interesting remains of ancient glass. (pp. 90, 91, 96.)

Woodston (1309). One mile south from Peterborough of which it forms a suburb. The church contains a small Saxon window. (pp. 47, 83, 94, 106, 115.)

Yaxley (1590). Four and a half miles south from Peterborough. A large village which formerly had a market. The church is large and handsome and has a fine crocketted spire; in the north transept there is a heart shrine, in which was found a heart, supposed to be that of William de Yaxley, Abbot of Thorney. In this parish is Norman Cross, where many of the French soldiers captured during the Napoleonic wars were imprisoned. (pp. 47, 55, 74, 76, 96, 115, 122.)

Yelling (242), six and a half miles north-east from St Neots, was anciently the property of the Lords Grey de Wilton.

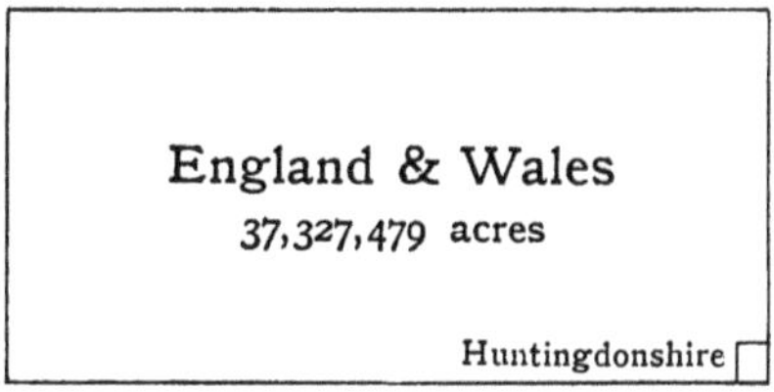

Fig. 1. Diagram showing the area of Huntingdonshire (234,218 acres) compared with that of England and Wales

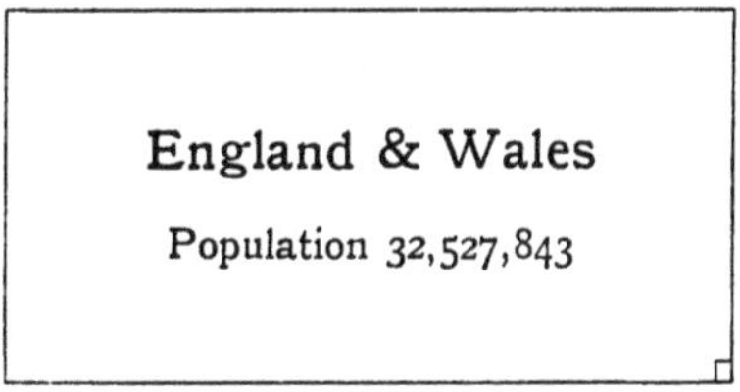

Fig. 2. Diagram showing the population of Huntingdonshire (57,771) compared with that of England and Wales

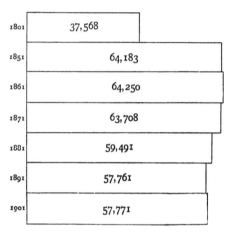

Fig. 3. Diagram showing changes in population of the ancient or geographical County of Huntingdonshire, from 1801—1901

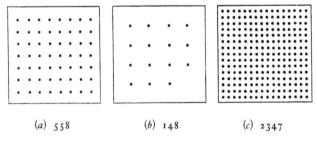

(a) 558   (b) 148   (c) 2347

Fig. 4. Diagram showing the density of Population to a square mile in (a) England and Wales, (b) Huntingdonshire, and (c) Lancashire.

(*Note, each dot represents* 10 *persons*)

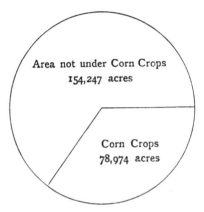

Fig. 5. Proportion of Cereal Crops to other Areas
in Huntingdonshire in 1908

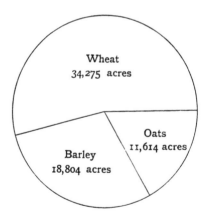

Fig. 6. Proportionate Area of Chief Cereals in
Huntingdonshire in 1908

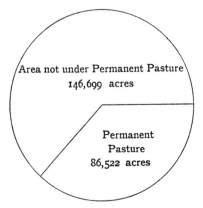

Fig. 7. Proportion of Permanent Pasture to other Areas
in Huntingdonshire in 1908

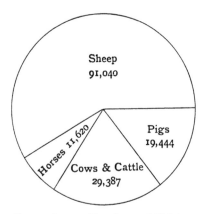

Fig. 8. Proportionate Numbers of Chief Live-stock
in Huntingdonshire in 1908

Milton Keynes UK
Ingram Content Group UK Ltd.
UKHW032321161024
449665UK00001B/14

9 781107 645967